Bad
Astronomy

Bad Astronomy

Misconceptions and Misuses Revealed, from Astrology to the Moon Landing "Hoax"

PHILIP C. PLAIT

John Wiley & Sons, Inc.

Copyright © 2002 by Philip C. Plait. All rights reserved.

Published by John Wiley & Sons, Inc., New York
Published simultaneously in Canada

Illustrations by Tina Cash Walsh

This publication is designed to provide accurate and authoritative information in regard to the subject matter covered. It is sold with the understanding that the publisher is not engaged in rendering professional services. If professional advice or other expert assistance is required, the services of a competent professional person should be sought.

Library of Congress Cataloging-in-Publication Data is available from the publisher.

ISBN 0-471-40976-6 (paper)

Printed in the United States of America

10 9 8 7 6 5

CONTENTS

INTRODUCTION

I love bad science fiction shows. *Angry Red Planet, Voyage to the Bottom of the Sea, UFO,* all those old TV shows and movies in black and white or living color. I grew up on them. I'd stay up late watching TV, sometimes long after my folks would normally let me. I remember clearly coming home from third grade and asking my mom for permission to watch *Lost in Space.* I *worshipped* that show, Robot, Dr. Smith, Jupiter 2, and all. I wanted to wear a velour, multicolored V neck sweater, I had a crush on Judy Robinson—the whole nine yards.

Sure, I liked the good ones too. *Five Million Years to Earth* and *The Day the Earth Stood Still* were favorites of mine back then, and they still are. But the important thing to me wasn't that they were good or bad, or even if they made sense—I remember an Italian flick about a voyage to Venus that might have been written by Salvador Dali on acid. What was important was that they had aliens and rocket ships.

I would spend long hours as a child pretending to ride a rocket to other planets. I always knew I'd be a scientist, and I was pretty sure I wanted to be an astronomer. Those movies didn't discourage me because of their bad science; they *inspired* me. I didn't care that it's silly to try to blast a conventional chemical rocket to another star, or that you can't hear sounds in space. All I cared about was *getting out there,* and if I could do it by watching ridiculous movies, then so be it. I would have given anything—everything— to be able to step on board a spaceship and be able to see a binary star up close, or cruise through a nebula, or go out through the plane of our Galaxy and see it hanging in the sky, faint, ominous,

1

luminous, against a velvet canvas of blackness so dark you can hardly convince yourself that your eyes are open.

Nowadays it would be a bit harder for me to give up everything to take such a ride. Maybe I would so my daughter could someday . . . but that day is not yet here. We're still stuck here on the Earth, more or less, and the only way we can see distant vistas is either vicariously through the eye of the telescope or through the eye of a movie director. One of those eyes, perhaps, is a bit more clearly focused than the other. Despite my childhood yearnings, as an adult I can wish that movies did a better job of portraying astronomy (and astronomers) to the public.

The movies may be inspiring, which is their most important job as far as I'm concerned, but there is a downside to the bad astronomy. It muddies the distinction between fantasy and science, between what is only pretend and what can really happen. Movies can portray the make-believe so realistically that the line gets blurred. It's fair to say that most people don't understand all that much about how space travel, for example, really works. Space travel is complicated, difficult, and relies heavily on unfamiliar physics.

Movies, however, make it look easy. Just get in your ship and go! All you have to do is watch out for the stray meteor shower or alien starship and everything should work out pretty well. Unfortunately, it doesn't work out that way in the real universe. If it did, we'd have colonies on Mars and the other planets by now. I've given talks to audiences about movies and astronomy, and the question almost always arises: why aren't we on the Moon now? Why haven't we built starships, or at least colonized the solar system? Sometimes these are honest questions, and sometimes they are asked with an edge of impatience, as if the people asking the question are concerned that the National Aeronautics and Space Administration engineers aren't as up to speed as Scotty from *Star Trek*. The film industry makes a big impression on people and, as the scenes play over and over again, they worm their way into our brains. Movies show space travel all the time, but they show it incorrectly, and so it doesn't surprise me that the majority of the viewing public has the wrong impression about how it really works.

If movies were the only purveyors of scientific inaccuracies, there would hardly be a problem. After all, it's their *job* to peddle fantasy. The problem is, it *doesn't* stop there. The news media's job is to report the facts clearly, with as much accuracy as possible. Unfortunately, this isn't always the case. In general, national media do a fine job; most TV networks, newspapers, and magazines have enough money to maintain at least a small staff of experienced science journalists who do a good job reporting the news. Local news is more often the culprit in misrepresenting science. Local reporters may be inexperienced in the technical jargon and tools of science, and so will sometimes write amazingly inaccurate copy. This is a real problem, with perhaps no easy solution, since many local news outlets simply cannot afford to keep as many reporters needed who are knowledgeable in the vast number of topics covered in the news.

Not that I am sidestepping national news. I remember vividly watching the *Today* show on NBC in 1994. The Space Shuttle was in orbit, and it was doing an experiment, dragging a large, circular shield behind it. The idea was that the disk would clear out particles in its wake like a snowplow pushing snow out of the way, leaving a cleared trail behind it. In the ultra-grade vacuum behind the wake shield experiments were being conducted that took advantage of such an environment.

Anchor Matt Lauer was reporting on this experiment, and when he was finished, Katie Couric and Bryant Gumbel both commented that it must have been hard for Lauer to read that copy. All three laughed, and Lauer admitted he didn't understand what he had just said. Think about that for a moment: three of America's most famous journalists, and they actually *laughed at their own ignorance in science*! How would this be different if, say, the report had been about Serbia, and they laughed at how none of them knew where it was?

Needless to say, I was pretty well steamed. That event is actually what started me down the road of discussing Bad Astronomy; I decided to take action when I realized that millions of people in the United States were getting their information from people who didn't understand even the simplest of scientific events. The report itself was accurate, and may have even been written by someone who fully well knew what the Space Shuttle was doing, but what the public saw was three respected journalists saying tacitly that it's okay to be ignorant about science.

It *isn't* okay. In fact, it's *dangerous* to be ignorant about science. Our lives and our livelihoods depend on it. No one can doubt the power of computers in today's world, computers that rely on physics to operate and improve their performance. Science is what makes our houses warm, our cars go, and our cell phones ring. Medical science progresses very rapidly, with new medicines, treatments, and preventions coming out almost daily. We *must* understand the science of medicine to be able to make informed decisions about our health. In the United States, hundreds of billions of dollars are spent every single year on science and technology, disciplines with which the typical voting citizen has not even a passing familiarity. *That's your money.* You should understand not only how it's being spent but also why.

Unfortunately, getting reliable science information isn't all that easy. Science misconceptions and errors are propagated by the media in all its forms. Unfortunately once again, the problem doesn't stop there.

Anyone who has gone outside on a clear, warm night and lain down on a blanket to watch the stars may know the deep joy of astronomy, but *understanding* astronomy is a different matter. Unfortunately, astronomy—and science in general—has been under attack lately. This isn't anything new, really, but the recent publicity has been a bit more obvious. From NASA budget cuts to state school boards that promote antiscience, the atmosphere is more hostile than it has been before. The modern consumer is bombarded by pseudoscience at every turn. Most newspapers in the country carry an astrology column, and some even have columns by self-proclaimed psychics, but precious few devote even a single page a week to a regular column about new scientific results. Conspiracy theories abound that twist and pervert simple science into ridiculous claims that are tissue-thin, yet are accepted wholesale by hordes of believers. The World Wide Web propagates these theories and a host of others at light speed around the world, making it even harder to distinguish between what is real and what is fantasy. In this atmosphere it's no wonder there's so much confusion about science.

Still, there's hope. Science may be on the rebound. The Discovery Channel started small, and many critics predicted it would fail. Yet, just a few years later, it is the most highly rated basic cable channel, and they charge dearly for advertising. Bill Nye the Science Guy teaches science on TV to kids in a fun and engaging way. Even adults can watch the show and get a kick out of it. The web deserves its due—one of the most popular sites on the web is not for a rock star, or a TV celebrity, or something steamy you wouldn't let your kids see. The website to which I am referring belongs to NASA. Yes, *the* NASA. Their home on the web is one of the most popular sites on the planet. When the Sojourner Mars probe landed on the Red Planet in 1997, their web site scored millions of hits, more than any other event in the history of the then-young web. Since then, the site has had almost a billion hits. When the Space Shuttle serviced the Hubble Space Telescope late in 1999, the NASA web site got a million hits in a single day. When the comet Shoemaker-Levy 9 smashed into Jupiter in 1994, the web nearly screeched to a halt due to the overwhelming amount of traffic as people tried to find pictures of the event from different observatories. Other science-based web sites report traffic similar to these examples as well.

The public not only likes science, it wants more. A survey of the reading public was made by newspapers, and they found that more people would read about science news, if it were offered, than about sports, finance, or the comics. When I give public lectures about results from Hubble, people barrage me with questions, and I usually wind up staying late answering more questions from people curious about the universe around them.

Despite their desire, a lot of people harbor some odd notions about astronomy. Come to think of it, it's probably *because* of that desire. If you want something enough, you'll take anything to fill that void. People have an innate curiosity about the universe; this is almost certainly a simple outcome of evolution. People who are curious are likely to explore, to learn, to discover. That's a pretty good survival trait.

But if they cannot get to a reliable source of information, they'll accept something less than reliable. People like the world to be mysterious, magical. It's more fun to believe that UFOs are aliens watching us than it is to find out that the overwhelming number of ET sightings are due to misinterpretations of common things in the sky.

The truth can be hard, and so sometimes it really is easier to believe in fiction. Other times, the tale has just enough of the ring of truth that you might not question it. Do we have seasons because Earth moves closer and then farther away from the Sun? Can you really see stars during the day from the bottom of a well?

Over the years I have found that people tend to have a lot of odd ideas about astronomy. Those ones I just mentioned are just a few examples of the host of misconceptions floating around in people's brains. Did I say "floating"? I mean *entrenched*. Like the movie scenes that ensconce themselves in our memories, misconceptions about astronomy—about any topic—take root in our minds and can be very difficult to weed out. As Cardinal Woosley said, quoted by Alistair Fraser on his Bad Science web site, "Be very, very careful what you put into that head, because you will never, ever get it out."

Far be it for me to disagree with His Eminence, but I think he's wrong. It *is* possible to yank that idea out and plant a healthier one. As a matter of fact, I think sometimes it's easier to do it that way. I have taught astronomy, and found that even an interested student can be easily overwhelmed in a classroom by a fire-hose emission of facts, numbers, dates, and even pictures relating to astronomy. There's so much to learn, and it can be hard to find a toehold.

However, if you start with something students already know, or *think* they know, that toehold is already there. Do you think we have seasons because the Earth's orbit is an ellipse, and so sometimes we're closer to the Sun than others? Okay, fine. Can you think of something else that might cause it? Well, what else do you know about the seasons? They're opposite in opposite hemispheres, right? Southern winter is northern summer, and vice versa. So what does that imply about our theory of what causes the seasons?

I won't give away the answer here; you'll find a whole chapter about it later in this book. But I hope you see my point. If you start with something already there in people's heads, you can work with it, play with it, make them think about it. Starting with a known misconception is a wonderful hook that captures people's thinking, and it can be fun and highly rewarding to think critically about these ideas. What do you know that you know wrong?

Some ideas are better than others. People remember movies, right? Then why not start there? In *Star Wars*, Han Solo dodges asteroids in the Millennium Falcon to escape Imperial fighters. In *Armageddon*, the Earth prepares for the impact of an asteroid a thousand miles across. In *Deep Impact*, a giant comet explodes over the Earth, causing nothing more than a beautiful fireworks display.

If you've seen these movies, these are scenes you'll remember. That makes them a great place to discuss real astronomy, and not the fantasy represented by the movies. You can find out what asteroids really are like; how easy it is to spot a big one and how hard it is to move one; and just why they're extraordinarily dangerous, even after you blow one up.

My parents may have thought I was wasting my time as a kid watching those bad science fiction movies. It turns out I was simply laying the groundwork for my life's work.

You *can* turn Bad Science into Good Science if you start in the right place.

This book is my way of starting in that place. We'll take a look at a whole lot of bad astronomy. Some of the examples will sound familiar, others likely won't. But they're all misconceptions I've run across, and they're all fun to talk about and even more fun to think about.

We'll uproot those brain weeds and plant healthy greenery yet.

PART I

⍦⍦⍦⍦⍦⍦

Bad Astronomy Begins at Home

There's an old joke about a family packing up to move. When their neighbors ask them why they are moving, they reply, "We heard that most fatal accidents happen within ten miles of home, so we're moving twenty miles away."

Sometimes I wish it were that easy. As a relatively new parent, I hear a lot about how much education my daughter gets at home. We teach her how to talk, read, do math, socialize, watch TV, argue with us, get her way, be petulant if she doesn't, and so forth. But often it's the things we don't *mean* to teach that stick. Kids are natural scientists. They watch, absorb information, repeat experiments, and their laboratory is their immediate neighborhood: home, parents, friends, television.

Unsurprisingly, not all the information they gather is accurate.

Astronomy may be the study of everything outside the Earth—that's not a bad definition—but *bad* astronomy starts at home. Why travel to some distant galaxy halfway across the observable universe when you can find examples of errant science right in your own fridge, or even in the bathroom? Science is a way of describing the universe, and the universe surely includes egg cartons and your toilet.

In the next few chapters we'll see how, like charity, bad astronomy begins in the home. Unfortunately, it doesn't stay there. You may try standing an egg on its end on the first day of spring at home, but classrooms and television reinforce this experiment as some sort of higher truth. You may not wonder where stuff goes when you flush it down the toilet, but which way that stuff spins as it drains becomes the topic of conversation at water coolers and bars everywhere. Even our very language is sprinkled with bad astronomy, from phrases like "meteoric rise" to "light years ahead."

With luck, though, and a sprinkle of critical thinking, we can plug up the drain of knowledge and topple the egg of ignorance.

I

The Yolk's on You: Egg Balancing and the Equinox

Consider the humble chicken egg.

Outside, its hard white calcium shell is mostly round and smooth. It might have small bumps on it, or even tiny ridges and waves, but its overall geometry is so well defined that we use the term "egg-shaped" when we see something similarly crafted. The very word "ovoid" comes from the Latin for "egg."

Inside, we have the white part of the egg—the technical term is *albumen*—and the yellow yolk. This goo is what is destined to become a chicken, if we let it. Usually we don't. Humans have all sorts of dastardly schemes planned for chicken eggs, from the simple act of cooking them to such bizarre practices as frying them on sidewalks to show how hot it is and using them to "decorate" houses on Halloween night.

But there is an even weirder ritual performed with the ovum of the *Gallus domesticus*. Every year, all across the United States and around the world, this ritual is performed at the beginning of the spring season. On or about March 21, schoolchildren, newspeople, and ordinary citizens take a chicken egg and try to stand it on end.

A nonscientific survey—conducted by me, by asking audience members when I give public talks and people I meet at parties or standing in line at grocery stores—shows that about one-half the population has either heard of this practice or tried it themselves.

That's roughly 130 million people in America alone, so it's certainly worth investigating.

If you've watched this ritual, or have tried it yourself, you know that it takes incredible patience, care, and stamina. It also takes luck, a flat surface, and a sprinkling of bad astronomy.

At first glance you might not expect astronomy to play any great role here. However, like the cultural rites of ancient peoples, it's the *timing* that's important. This ritual is performed on the date of the spring equinox, which is the time when the Sun crosses from the southern to the northern hemisphere of the sky. The spring equinox is called the *vernal equinox* by astronomers; the root of the word "vernal" means "green," which has obvious links to springtime. To my mind, the idea of balancing an egg is as strange a way to celebrate the spring equinox as is dancing at the foot of Stonehenge dressed as Druids.

So what are the details of egg standing, exactly? It goes something like this: According to the legend, it's only possible to stand an egg on end and have it balance perfectly on the exact date of the spring equinox. Some people even claim that it must be done on the exact *time* of the equinox. If you try it any other time, even minutes before or after, you'll fail.

That's all there is to it. Seems simple, right? Every year at the magical date, newscasters—usually TV weatherpeople, since the date has climatological ramifications—talk on the air about balancing eggs. A lot of schoolrooms, in an effort to perform a scientific experiment, also try to get the little ova upright. Sometimes the newscasters will go to the classroom to show the tykes trying, and after a while, voilà! Someone gets an egg to stand! The cameraman is rushed over and the beaming future scientist gets his or her face on TV that night, film at eleven.

Unfortunately, if the teacher doesn't go any further, the child's future as a scientist may be in some doubt. This hasn't really proven the legend one way or another. Let's take a closer look at it.

We need to start by asking what should be an obvious question: why would the vernal equinox be the only time you can do this? I have asked that of people who believe the legend to be true, and they make vague claims about gravity aligning just right on that special day. The Earth, the egg, and the Sun all line up just right to let the egg balance. But this can't be right: there is always some point on the surface of the Earth exactly between the center of the Earth and the Sun. It has nothing to do with any special time. And shouldn't the Moon have some effect too? The Moon's gravitational force on the Earth is pretty large, so its gravity is pretty influential. Yet the Moon plays no part at all in the legend. Obviously, the vernal equinox is not the root of the issue.

Luckily, we don't have to rely totally on theory. The legend of vernal egg-balancing makes a practical prediction that can be tested. Specifically, the prediction is: *If an egg will stand only on the vernal equinox, then it will not stand at any other time.* Once you think of it that way, the experimental verification is obvious: try to stand an egg on end some other time. The vernal equinox is usually on March 21 or thereabouts every year. To test the theory, we need to try to upend an egg on some other day, a week, month, or even farther from the time of the equinox. The problem is, most people don't follow through with the experiment to its logical conclusion. They only try it on the equinox, and never on any other day.

However, I've tested it myself. The picture shows not just one but *seven* eggs standing on end in my kitchen. Of course, you're skeptical—as you should be! Skepticism is an important scientific tool. But why take my word for it? Chances are it's not March 21 as you read this. Go find some eggs and give it a try. I'll wait.

❧ ❧ ❧

Finished? So, could you do it? Maybe not. It's not easy, after all. You need patience, a steady hand, and a fairly strong desire to balance an egg. After I got those eggs balanced, I had trouble balancing any more. My wife happened to come downstairs at that moment and asked me what the heck I was doing, and she quickly decided that it looked like fun. Actually, I think it was her competitive nature that drove her; she wanted to stand up more eggs than I did. She did. Actually, she had a hard time at first. I told her that

Standing eggs on end has nothing to do with the time of year, and everything to do with a steady hand, a bumpy egg, and lots of patience. These eggs were photographed in autumn, months after the vernal equinox. (But don't take my word for it; try it yourself.)

I had heard it's easier to stand an egg if you shake it a little first to help the yolk settle. She did, but pressed too hard on the shell. While she was shaking it her thumb broke through the shell, and she got glop all over the wall of our kitchen! I imagine we have the only house in the country where something like this could happen.

Eventually, she was successful. She was the one who got the rest of the eggs to stand; we got eight total from that one carton. Clearly, her hands are steadier than mine. Once, when scheduled to give a public talk about Bad Astronomy at the Berkshire Museum of Natural Science in Pittsfield, Massachusetts, I arrived late due to an ice storm. I had to change my clothes quickly and literally run to the auditorium. When I arrived, I was out of breath and my hands were shaking a little from the stress and excitement. I usually start off the lecture by balancing an egg, but because I was shaking a little I had a very hard time of it! I struggled with the eggs all during the time the lecture series curator was introducing me, and by some sort of miracle I got it balanced just as he finished announcing my name. To this day, it's the loudest and most pleasing ovation I have ever received.

The lesson here is that if you are patient and careful, you can usually get one or two eggs from a carton to stand. Of course, you can also cheat. If you sprinkle salt on the table first, it will support the egg. Then you gently blow on the remaining salt so that it gets swept away. The salt holding up the egg is almost invisible, and will never be seen from a distance. I, however, would never do something like this. Honest! Actually, over the years I have become pretty good at balancing eggs with no tricks. Practice makes perfect.

Still, this doesn't answer the question of *how* an egg can balance at all. It's such an odd shape, and oddly balanced. You'd really just expect it to fall over every time. So just why does an egg stand? I'll admit to some ignorance of the structure of eggs, so to find out more about it I decided to find an expert.

I found a good one right away. Dr. David Swayne is a poultry veterinarian for the United States Department of Agriculture in Athens, Georgia. When pressed, he admits to knowing quite a bit about chicken eggs. I bombarded him with questions, trying to get to the bottom, so to speak, of the anatomy of an egg. I was hoping that somewhere in the structure of an egg itself was the key to balancing them (although I forgot to ask him which came first, it or the chicken).

The characteristic shape of an egg, he explained to me, is due to pressure from the chicken's reproductive system as the egg is pushed through the reproductive organs. The yolk is made in the ovary, and the albumen is added as the yolk is forced through a funnel-shaped organ called an infindibulum. The white-yellow combination is only semi-gooey at this point, and it is covered with a membrane. The infindibulum forces the egg through using peristalsis, a rhythmic squeezing and relaxing of the infindibulum. The back part of the egg getting pushed gets tapered from being squeezed, and the end facing forward gets flattened a bit. That's why an egg is asymmetric! Eventually, the egg reaches the shell gland, where it sits for roughly 20 hours and has calcium carbonate deposited all around it. That's what forms the shell. The calcium carbonate comes out in little lumps called concretions, which is why eggs sometimes have little bumps on the bottom. Once the shell is formed, the egg goes on its way out the chicken. (At this point I'll stop the narrative and you can use your imagination for the last part of the egg's journey. After hearing Dr. Swayne discuss it I couldn't eat an omelet for weeks.)

At this point I had two theories about egg balancing. One was that if you let the egg warm up, the albumen will thin a bit and the yolk will settle. Since the yolk moves down, the center of gravity of the egg lowers, making it easier to stand. Dr. Swayne put that to rest pretty quickly. "The viscosity of the albumen doesn't depend on temperature," he told me. "It's designed to keep the yolk pretty much in the middle of the egg." That makes sense; the yolk is the embryo's food and shouldn't get jostled too much. The albumen keeps it from bumping up against the inside wall of the shell, where it might get damaged. A thinned albumen can't do its job, so it has to stay thick. Warming the eggs won't help much in standing them on end.

My other working theory relied pretty heavily on those little calcium carbonate bumps. They are almost always on the bottom, fatter end of the egg. According to my theory, these imperfections act like little stool legs, which help support the egg. Through my own experimenting I found that a smooth egg is very difficult if not impossible to stand up, but a bumpy one is actually pretty easy, once you get the hang of it. So it's not the vastness of space and the infinite subtle timings of the Earth as it orbits the Sun that gets the egg to balance, I concluded, it's the stubby little bumps on the end. So much for the grandeur of science.

Yet the legend persists. Science and reason are a good arsenal to have in the battle against pseudoscience, but in most cases they take a backseat to history and tradition. The egg-balancing legend has been around for a while, and is fairly well ensconced in the American psyche. I get lots of e-mail from people about standing eggs on end, especially around the middle of March, shortly before the equinox. A lot of it is from people who think I am dead wrong. Of *course* it's all about the equinox, they telwl me. Everyone says so. Then they tried it on the day of the equinox for themselves, and it worked! The egg stood!

Of *course* it did, I tell them. It'll stand on any other day as well, which they can prove to themselves if they just try it. They haven't followed through with their experiment, and they convince themselves they are right when the evidence isn't all in. They rely on word of mouth for what they believe, and that isn't a very strong chain of support. Just because someone says it's so doesn't make it so. Who knows from where he or she first heard it?

In this case, we can find out. Most urban legends in America like this one have origins that are lost in the murky history of re-peated tellings. However, happily, this one has a traceable and very specific origin: *Life* magazine. As reported by renowned skeptic Martin Gardner in the May/June 1996 issue of the wonderfully rational magazine *Skeptical Inquirer*, the legend was born when, in the March 19, 1945 issue of *Life,* Annalee Jacoby wrote about a Chinese ritual. In China, the first day of spring is called Li Chun, and they reckon it to be roughly six weeks before the vernal equi-nox. In most countries, the equinoxes and solstices do not mark the beginning of seasons; instead, they're actually the midpoints. Since a season is three months or twelve weeks long, these coun-tries believe that the actual first day of spring is six weeks before the equinox.

The Chinese legend has an uncertain origin, according to Mr. Gardner, although it is propagated through old books about Chi-nese rituals. In 1945 a large number of people turned up in the city of Chunking to balance eggs, and it was this event that Ms. Jacoby reported to *Life.* Evidently, the United Press picked up the story and promptly sent it out to a large number of venues.

A legend was born.

Interestingly, Ms. Jacoby reported that balancing an egg was done on the first day of spring, yet it was never said—or else it was conveniently forgotten—that the first day of spring in China was a month and half before the first day of spring as recognized by Americans. This inconvenient fact should have put a wet blan-ket on the proceedings, but somehow that never slowed anything down.

In 1983 the legend got perhaps its most famous publicity. Donna Henes, a self-proclaimed "artist and ritual-maker," gathered about a hundred people in New York City to publicly stand eggs up at the exact moment of the vernal equinox on March 20, 1983. This event was covered by the *New Yorker* magazine, and a story about it appeared in its April 4, 1983 issue describing how Ms. Henes handed out eggs to the onlookers, making them promise not to stand any up before the appointed time. Around 11:39 P.M. she upended an egg and announced, "Spring is here."

"Everyone in the crowd, us included, got busy balancing eggs," the *New Yorker* effused. "Honest to God, it works." The unnamed reporter was not so convinced, however, as to swallow this line whole. Two days after the equinox, the reporter brought a dozen eggs to the same place where the ritual had occurred. For twenty minutes the reporter tried to balance the eggs but didn't get a single one to stand on its end.

The reporter admits the failure may have been psychological. "The trouble may have been that we didn't want the egg to balance—that we wished to see Donna Henes to be proved right." This, despite the reporter having asked several physicists about the legend, and having all of them say they couldn't think of why it should work. I find it ironic and faintly troubling that one of those physicists said that water swirls down the washbasin drain one way in the northern hemisphere and the other way in the southern hemisphere—this is another astronomy-based urban legend, and it is not true. (See chapter 2, "Flushed with Embarrassment," for more on that.)

Ms. Henes went on to more balancing rituals, too. The year after the 1983 demonstration, more than 5,000 people showed up at the World Trade Center to participate in an egg balancing. Even the *New York Times* was duped; four years later, on March 19, 1988, they published an editorial with the headline: "It's Spring, Go Balance an Egg." Two days later, the *Times* ran a picture of people standing eggs up once again at the World Trade Center.

So this legend seems to spread easily. If the illustrious *New York Times* can help it along, the transmission may very well be unstoppable. Still, stoppable or not, I cannot let something like this get past me so easily. In an effort to stem the tide, just before the vernal equinox of 1998, I called a local TV station and chatted with the weatherman about egg standing. He had never heard of it, but was excited because they like to have little quizzes before the forecast, and this was a good topic for the news on the equinox. So he asked the rest of the news team, composed of two anchors and a sportscaster, if an egg could stand only on the equinox. Who do you think got it right? To my surprise, the sportscaster figured the equinox had nothing to do with it, while the two news anchors both guessed it did. It's funny, too; the anchors never did get their eggs to stand up, while the sportscaster did. A triumph for science!

It may simply be that our common sense—something short and round like an egg can't stand on end—and poor recall—who can remember exactly *why* we have seasons?—combine to reinforce the legend. Worse, it gets positive feedback from the newscasts every year. Not every TV news station is as open-minded as the one I called. Imagine the real schoolroom scene described at the beginning of this chapter. We have 30 or so kids and one harried teacher, going from student to student giving encouragement. Suddenly, a child gets an egg to stand. At the same time, 29 other kids *don't* get theirs to stand. Who gets on TV? Right. It's no fun to show the ones who didn't get it. However, science isn't just about showing when you're right; it's also about showing when you're wrong.

A lot of my mail is also from people who *did* follow through. I received an e-mail from Lisa Vincent, who teaches at Mancelona Middle School in Mancelona, Michigan. She decided to test the egg myth for herself, and had her students try it on October 16, 1999, which, incidentally, is almost one year after the photos of my own test were taken (see page 14). Not only were Ms. Vincent and her students able to balance several eggs five months before the vernal equinox but they were also able to balance the eggs on their small ends. For proof she sent me a photograph of her proud students and their eggs standing in what looks to me like an upside-down position. That is a feat I had never been able to accomplish up until then, and I must admit a tinge of professional jealousy. I had always assumed it couldn't be done. However, after knowing it *could* be done, I tried even harder, and eventually managed to upend an egg on its narrow tip. It just goes to show you, even scientists need to have their world rocked on occasion.

Incidentally, Ms. Vincent told me that the eggs stood balanced that way until she decided to take them down on November 21, over a month after they were placed there. Here we have a great example of people not being willing to accept what they hear, and wanting to try it for themselves. That is the essence of science.

The essence of science is that it makes its own improvements: A theory is only as good as its next prediction. Remember my own theory about the stubby bumps supporting the eggs? Well, Ms. Vincent's middle-school class showed me I was wrong. They balanced the eggs on their tops, and I have never seen a top of an egg that wasn't smooth. Certainly, the bumps make it easier since I am always able to balance bumpier eggs more easily than smoother ones. But the bumps must not be *critical* to balancing, or else the eggs wouldn't balance on their short ends. Clearly, these kids balanced the eggs through perseverance and strong desire. One of the beauties of science is that it improves itself, and another is that you never know where that improvement will come from. Mine came from Mancelona, Michigan.

Science is about asking, why? and, why not *this* way? Sometimes you need to think *around* the problem. For example, if the spring equinox is special, isn't the autumnal one special, too? They are both basically the same, yet you never hear about people trying to stand eggs on end in September. Even better, the seasons are *opposite* in the northern and southern hemispheres; when it's spring in one it's fall in the other and vice versa. But people usually don't think of these things. It's too easy to simply accept what you're told. This is extraordinarily dangerous. If you just assume without thinking critically that someone is right, you may be voting for the wrong politician, or accepting a doctrine that has a bad premise, or buying a used car that might kill you. Science is a way of distinguishing good data from bad.

Practicing science is wonderful. It makes you *think* about things, and thinking is one of the best things you can do.

2

❧❧❧❧❧❧

Flushed with Embarrassment: The Coriolis Effect and Your Bathroom

I t's a pretty scene. Nanyuki is a small town situated just north of the equator where it cuts across Kenya in Africa. The town was founded early in the twentieth century, and still has something of a frontier feel to it.

It's a frequent stop for tour buses on their way to nearby Mount Kenya. It has the seemingly mandatory gift and curio shops, but it also features a local man named Peter McLeary. As tourists gather around, McLeary shows them a demonstration they are not likely to forget. More's the pity.

McLeary takes the tourists to a line drawn on the floor of an old burned-out hotel, and tells them that it's the actual location of the equator. A glib speaker, he explains that water swirls down a drain clockwise north of this line and counterclockwise south of it, an effect caused by the rotation of the Earth.

He then goes on to prove it. He takes a small, roughly square pan about 30 centimeters across and fills it with water. He places some matchsticks in it so that his audience can see the rotation more easily. Walking to one side of the line and turning to face his audience, he pulls out a stopper, letting the water drain out. Sure enough, when he does this demonstration north of the equatorial line the water drains clockwise, and when he repeats the experiment south of the line, the water drains the other way. Proof positive that the Earth is spinning!

The demonstration is convincing, and McLeary has done it for many years, raking in tips from the credulous tourists. It has been seen by countless travelers, and was even featured on the PBS series *Pole to Pole,* in which former *Monty Python* silly man Michael Palin tours the world, taking in interesting sights. In this particular episode Palin watches McLeary do his thing and adds, "This is known as the Coriolis effect . . . it *does* work."

Actually, no, it doesn't. Palin, and who knows how many tourists before and after him, are being taken in by a fraud. And it doesn't end there. This hoary idea is used to explain why toilets flush in different directions in the northern and southern hemispheres, as well as the way northern and southern sinks and bathtubs drain. Many college students claim that their high school science teachers taught them this fact. The problem is, it's no fact. It's bad astronomy.

The Coriolis effect is real enough. By the 1800s, it had been known for years that cannonballs fired along a north-south line tended to deviate from a straight path, always landing west of their mark if fired toward the south, and east if fired to the north. In 1835 the French mathematician Gustave-Gaspard Coriolis published a paper with the unassuming title of, "On the Equations of Relative Motion of Systems of Bodies." In it, he describes what has become known as the Coriolis effect.

Imagine you are standing on the Earth. Okay, that's easy enough. Now imagine that the Earth is spinning, once a day. Still with me? Okay, now imagine you are standing on the equator. The rotation of Earth takes you eastward, and after a day you have swept around a big circle in space, with a radius equal to the Earth's radius. On the equator, that means you have traveled about 40,000 kilometers (25,000 miles) in one day.

Now imagine you are on the north pole. After one day, you have rotated around the spot on which you are standing, but you haven't actually *gone* anywhere. The north pole is defined as the spot where the Earth's rotation axis intersects the ground, so pretty much by definition you don't make a circle there. You just spin, making no eastward motion at all.

As you move north from the equator, you can see that your eastward velocity decreases. At the equator you are moving nearly 1,670 kilometers per hour (1,030 miles per hour) to the east (40,000 kilometers in 24 hours = 1670 kph). At Sarasota, Florida, at a latitude of about 27 degrees north, you are moving east at 1,500 kph (930 mph), and by the time you reach Wiscasset, Maine, at 44 degrees north latitude, you are moving east at only 1,200 kph (720 mph). If you brave the chill of Barrow, Alaska, you'll be at latitude 71 degrees north and moving at a leisurely 550 kph (340 mph). Finally, at the north pole, you aren't moving east at all; you just make your tiny circle without any eastward movement.

Let's say you stop in Sarasota, which is a reasonable thing to do, given the climate there compared to Barrow. Now imagine someone on the equator due south of your position takes a baseball and throws it directly north, right toward you. As it moves northward, its velocity *eastward* increases relative to the ground. Relative to you, that baseball is moving 1,670 kph − 1,500 kph = 170 kph (1,030 mph − 930 mph = 100 mph) or so to the east by the time it reaches you. Even though the fastball is aimed right at you, it will miss you by a pretty wide margin! By the time it gets to your latitude, it will be a long way to the east of you.

That's why cannonballs are deflected as they travel north or south. When they are first shot from the cannon, they have some initial velocity to the east. But if they are fired north, they reach their target moving faster to the east than the ground beneath them. The cannoneer needs to aim a bit west to compensate. The reverse is true if it is fired south; the cannonball reaches its target moving *slower* than the ground, and needs to be aimed to the east to actually hit the target.

In our baseball example above, the distances and times involved were large, letting the Coriolis effect gather some steam. In reality, it's a tiny effect. Let's say you are driving a car north at 100 kph (60 mph) in Wiscasset, Maine. The Coriolis effect deflects you by the teeny amount of 3 millimeters (0.1 inches) per second.

After a solid hour of driving, that amounts to a deflection of only 10 meters (33 feet). You couldn't possibly notice this.

Still, it *is* there. It's subtle, but over long distances and large amounts of time it adds up. That can be a mighty sum, given the correct circumstances.

And those circumstances do arise. An area of low pressure in the atmosphere is like a vacuum cleaner, drawing in the surrounding air. Let's take the simplified view that we are in the northern hemisphere, and assume that the air is coming in only from due north and due south. The air coming in from the south is moving faster to the east than the air near the center of the low-pressure system, so it bends to the east. Air moving from the north is moving slower than the air in the center of the system, and deflects west. These two deflections add up to a counterclockwise rotation to the low-pressure system. This is called a **cyclonic** system.

The opposite is true in the southern hemisphere. A low-pressure system will spin clockwise because air drawn in from the north will be moving faster to the east, and air coming in from the south will be moving slower. The spin is opposite from the northern hemisphere.

If the system is stable for a long time, days or weeks, it can grow massively in strength. Warm ocean water feeds the system, making it stronger. As the air gets closer to the center it moves faster, like an ice skater who spins faster when she draws in her arms. If the winds can gain in strength and blow at a hundred or more kilometers per hour, it becomes a hurricane (or a typhoon if it's in the Pacific ocean).

All that, from that tiny deflection you can't even feel in a car!

Does this sound familiar? Sure! It's the same idea that Peter McLeary uses to explain why water swirls the way it does when he gives his demonstration in Kenya.

But there's a problem: as we already saw, the Coriolis effect only produces a measurable effect over huge distances and long periods of time. Even the most decadent of bathtubs is thousands of times too small and drains way too quickly to ever be affected by it. It can be shown mathematically that random motions in your water are thousands of times stronger than the Coriolis effect,

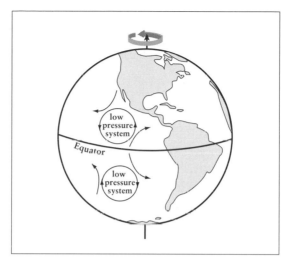

The Coriolis effect is only significant over large distances. A hurricane is born when a low-pressure patch of air draws air in from higher and lower latitudes. Because of the Coriolis effect, in the northern hemisphere the air from the south moves east, and the air from the north moves west, causing a clockwise rotation.

which means that any random eddy or swirl in the water will completely swamp it. If the water always drains one way from your bathtub, then it has far more to do with the detailed shape of your drain than from the rotating Earth.

Obsessive would-be physicists have actually performed experiments using household sinks. They have found that the sink needs to sit still for over *three weeks* so that random currents die off enough to see an appreciable Coriolis effect. Not only that, they have to let the sink drain one drip at a time to give the effect time to take hold. You're not likely to see this after hand-washing your delicates in the sink.

The same is true for your toilet. This one always makes me laugh: toilets are *designed* to spin the water. It helps remove, well, stubborn things that don't want to be removed so easily. The water is injected into the bowl through tubes that are angled, so it always flushes the same way! If I were to rip my toilet out of the wall and fly it down to Australia, it would flush in the same direction it does now.

The idea that the Coriolis effect works on such small scales is a pernicious myth. I have seen it in countless television shows and magazine articles; it was once even reported in the *Sports Illustrated* swimsuit issue. Oddly, they describe walking across the equator from the Central American country of Costa Rica, which is hundreds of kilometers from the equator. Some writer on staff did the figures incorrectly, but then, those aren't the kind of figures the magazine is usually trying to sell. On the other hand, maybe all that walking is how the models stay so slim.

So, if the Coriolis effect doesn't work on something as small as a sink or a pan, how did Peter McLeary pull it off? After all, as Michael Palin commented, it worked for him.

Actually, McLeary cheated. If you watch him do it on *Pole to Pole*, you can catch the swindle. He stands on his equator line and fills the basin. Then he walks a few meters or so north, and *rapidly turns to his right to face his audience*. He opens a hole in the bottom of the pan and the water obligingly rotates clockwise as it drains out. Next, he refills it, walks a few meters south of the equator, then *rapidly turns to his left to face the audience*. Draining, the water spins counterclockwise.

Do you see how this works? By spinning rapidly in opposite directions, he can make the water rotate any way he wants! The squarish shape of the pan helps, too; the corners help push on the water as the pan rotates, making it flow better.

Meteorology professor Alistair Fraser has used this demonstration in his own class. He draws a line down the middle of the classroom and declares it to be the equator (he teaches in Pennsylvania). He then does just what McLeary does and gets the same results.

Still don't believe me? Then think about it: the Coriolis effect should make draining water spin counterclockwise in the northern hemisphere and clockwise in the southern. In the northern hemisphere, water moving north deflects east, moving it counterclockwise. Water coming south from the north deflects west, but that's still counterclockwise. The opposite is true again for the southern hemisphere; the water will spin clockwise.

But this is precisely the *opposite* of what McLeary demonstrates. He's a fraud!

Your honor, I rest my case.

Well, not really. I have one more tale to tell. While searching for information about Nanyuki, I found one tourist's travelogue that describes three sinks sitting roughly ten meters apart, just outside of town. One is south of the equator, the second is directly on it, and the third is north of it. Perhaps someone else is horning in on McLeary's act. Anyway, the tourist who wrote the travelogue claimed that the northern sink drained clockwise, the southern sink drained counterclockwise, and the one in the middle drained straight down. Evidently the drain holes have been cut in such a way as to force the water to drain the way the designer wanted. Note once again that they drain the wrong way!

It's pretty funny, actually. They go through all that trouble to make a few bucks, and they don't even get the scam right. Somehow, though, I don't think those con artists are starving. Con artists rarely do. They can always put the right spin on their subjects.

3

✄ ✄ ✄ ✄ ✄ ✄

Idiom's Delight: Bad Astronomy in Everyday Language

LIGHT-YEARS AHEAD

One of the reasons I loved astronomy when I was a kid was because of the big numbers involved. Even the nearest astronomical object, the Moon, was 400,000 kilometers away! I would cloister myself in my room with a pencil and paper, and painstakingly convert that number into all kinds of different units like feet, inches, centimeters, and millimeters. It was fun, even though it branded me as a geek. That's all changed, of course. As an adult I use a computer to be a geek a million times faster than I ever could when I was a kid.

The fun really was in the big numbers. Unfortunately, the numbers get too big too fast. Venus, the nearest planet to the Earth, never gets closer than 42 million kilometers from us. The Sun is 150,000,000 (150 million) kilometers away on an average day, and Pluto is about 6,000,000,000 (6 billion) kilometers away. The nearest star to the Sun that we know of, Proxima Centauri, is a whopping 40,000,000,000,000 (40 trillion) kilometers away! Try converting *that* to centimeters. You'll need a lot of zeros.

There is a way around using such unwieldy numbers. Compare these two measurements: (1) I am 17,780,000,000 Angstroms tall. (2) I am 1.78 meters tall. Clearly (2) is a much better way to express my height. An Angstrom is a truly dinky unit: 100 million of them would fit across a single centimeter. Angstroms are used to measure the sizes of atoms and the wavelengths of light, and they are too awkward to use for anything else.

The point is that you can make things easy on yourself if you change your unit to something appropriate for the distances involved. In astronomy there aren't too many units that big! But there is one that's pretty convenient. Light! Light travels *very* fast, so fast that no one could accurately measure its speed until the nineteenth century. We now know it travels about 300,000 kilometers every second. That's a million times the speed of sound! No wonder no one could measure it until recently.

So, astronomers use light itself as a big unit. It took the Apollo astronauts 3 days to go to the Moon in their slowpoke capsule, but it takes a beam of light just 1.3 seconds to zip through the same trip. So we say the Moon is 1.3 **light-seconds** away. Light takes 8 minutes to reach the Sun; the Sun is 8 light-minutes away. Distant Pluto is about 6 light-hours away.

A light-minute or -hour may be useful for solar system work, but it's small potatoes on the scale of our Galaxy. Light doesn't travel far enough in only one minute. For galactic work, you need a **light-year**, the distance a beam of light travels in one year. It's equal to about 10 trillion kilometers, which is a long way. Proxima Centauri is 4.2 light-years away; the light leaving a presidential inauguration might not reach Proxima Centauri until after the president leaves office at the end of the term!

The light-year is the standard yardstick of astronomers. The problem is that pesky word "year." If you're not familiar with the term, you might think it's a time unit like an hour or a day. Worse, since it's an astronomical term, people think it's a really long time, like it's a lot of years. It isn't. It's a *distance*.

That doesn't stop its misuse. The phrase "light-years ahead" is a common advertising slogan used to represent how advanced a product is, as if it's way ahead of its time.

I can picture some advertising executive meeting with his team, telling them that saying their product is "years more advanced than the competition" just doesn't cut it. One member of the ad team timidly raises a hand and says, "How about if we say 'light-years' instead?"

It sounds good, I'll admit. But it's wrong. And more bad astronomy is born.

Worse, one Internet service provider even claims it's "light-years **faster** than a regular connection." They're using it as a *speed*!

Not surprisingly, Hollywood is a real offender here. In the first *Star Wars* movie, for example, Han Solo brags to Obi Wan Kenobi and Luke Skywalker that he could make the Kessel Run in "less than twelve parsecs." Like a light-year, a parsec is another unit of distance used by astronomers; it's equal to 3.26 light-years (that may sound like a silly unit, but it's actually based on an angular measure using the size of the Earth's orbit). Han's claim is like runners saying that they run a 10-kilometer race in 8 kilometers! It doesn't make sense. Astute fans of *Star Wars* may notice that Obi Wan gets a pained look on his face when Han says that line. Maybe he is wincing at his pilot's braggadocio; I choose to think Obi Wan knows his units.

METEORIC RISE

If you go far from the lights of a city on a clear night and wait long enough, chances are you'll see a shooting star. The proper name for it is a **meteor.** Of course, meteors aren't stars at all. They are tiny bits of gravel or dust that have evaporated off the surface of comets during their long voyages around the Sun. Some are the shrapnel from collisions between asteroids. Most of them are very small; an average one is about the size of a grain of sand.

While they are out in space, these specks are called **meteoroids.** They orbit the Sun as the Earth does, and sometimes their paths cross ours. When one does, the little piece of flotsam enters our atmosphere, and the tremendous pressure generated by its travel through our air causes it to heat up tremendously, so hot that it glows. That glow is what we call a meteor. If it impacts the ground, it's called a **meteorite.**

These three names cause a lot of confusion. A meteoroid glows as a meteor when it moves through the air, and it becomes a mete-orite when it hits the ground. I got into an argument once with a friend about what to call meteors during various parts of their travel. I said they are meteorites when they hit the ground. He asked, "What if they hit a house and stop on the second floor?" I countered that the house is in direct contact with the Earth, so it's still a meteorite. He rebutted by asking, "What if it hit an airplane and stopped?"

I had to scratch my head over that one. Is it a meteorite when the plane lands? What if the plane crashes? At this point we decided we were being silly, and decided to just go outside and look for meteors. That may have saved our friendship.

Anyway, meteors start off in space and then fall to the Earth. They appear dramatically, flashing into our view, and burn out suddenly as they descend through the atmosphere toward the ground, sometimes leaving a long trail of glowing ash behind them. They start off bright, then fade away.

Enter bad astronomy. I was reading a major metropolitan newspaper one day and was amused when it referred to a Russian official's "meteoric rise" in the political structure of that country. Of course, the reporter meant that the official appeared out of nowhere and made a quick, brilliant rise to the top of his heap. The *real* meaning of the phrase, however, is just the opposite: were we to be literal, the official would have made a sudden, eye-catching appearance in the political arena and then quickly burned himself out as he descended the ranks. He may have left a trail behind him, and even made quite an impact in the end!

DARK SIDE OF THE MOON

I had the misfortune one morning to wake up to the radio playing the song "Dream Weaver." I'll admit I used to love that song when I was a kid, but as a friend of mine likes to say, "We are not responsible for songs we liked when we were 15 years old." Anyway, as the tired, hackneyed verses went on, one in particular caught my ear: "Fly me away to the bright side of the moon, and meet me on the other side."

Of course, there is a bright side of the Moon, and you can go to it. But if you sit still, you can only be there for two weeks, max.

The bright side, and therefore the dark side as well, is not a fixed place, but appears to move as the Moon rotates.

Seen from the surface of the Earth, the Moon does not appear to rotate. It seems to show the same face to us all the time. Actually it does spin; it's just that it spins once for every time it goes around the Earth. Its rotation teams up with its revolution in such a way that it always shows that one face to us. We call that face the **near side** of the Moon. The other side, the one we never see, is called the **far side.** The far side of the Moon has only been seen by probes or by astronauts who have actually orbited the Moon. Since it's remote and not well known, the far side of the Moon has become synonymous with something terribly far away or unexplored.

The problem is, people confuse the *far* side with the *dark* side. You almost never hear the phrase "far side of the Moon." It's always the "dark side of the Moon." This phrase isn't really wrong, but it *is* inaccurate.

Like the Earth, the Moon spins. The Earth spins once every 24 hours, so that someone standing on its surface sees the Sun go up and down once a day. As seen from outside the Earth, that person is on the dark side of the Earth when he or she is on the half that is facing away from the Sun. But the dark side is not a permanent feature! Wait a few hours, and the Earth spins enough to bring that person back into the sunlight. He or she is now on the bright side of the Earth. No part of the Earth is on the dark side forever.

The same goes for the Moon, except its day is 29 of our Earth-days long. Someone on the Moon will see the sun set two weeks after it rises! Since half the Moon is in sunlight and half in darkness, there *is* technically a dark side to the Moon, but it changes as the Moon rotates. Except near the poles, a single point on the Moon is in sunlight, then in darkness, for two weeks.

You can see that the dark side of the Moon is simply just the *night* side of the Moon. It is no more a fixed feature than the night side of the Earth. Sometimes the far side *is* the dark side, but it's also sometimes the bright side. It just depends on when you look.

One of the best selling music albums of all time is Pink Floyd's *Dark Side of the Moon.* It may be popular, but astronomically it's in eclipse.

Incidentally, at the end of that album there is a quiet voice-over: "There is no dark side of the moon. As a matter of fact, it's all dark." In a sense that line is correct: the Moon is actually very dark, only reflecting less than 10 percent of the sunlight that hits it. That makes it about as dark as slate! The reason it looks so bright is that it is in full sunlight, and that means there's a lot of light hitting it. Ironically, even though six Apollo missions landed on the near side of the Moon, they only explored the tiniest fraction of the surface. In essence, even the *near* side of the Moon is largely unexplored, and it's still very far away.

Now, to be honest, there may be a part of the Moon that's always dark. Near the poles there are deep craters with raised rims around them. From that region the Sun is always near the horizon, just like at the poles on Earth. Since the craters on the Moon can be deep, the Sun may always be hidden by the rim of the crater. Sunlight never reaches the bottom of such craters! There is tantalizing evidence of ice at the bottom of such craters, untouched by the warming rays of the Sun. If it's true, there are two major implications. One is that the ice can be used by lunar colonists for air and water, negating the need to carry it along with them from Earth. That saves a vast amount of money, fuel, and effort.

The other implication is that the phrase "dark side of the Moon" actually has a limited truth to it—as far as the dark crater bottoms go! Maybe I need to start a "Not-So-Bad Astronomy" web site.

QUANTUM LEAP

Sometimes the advertising executives we discussed earlier aren't satisfied with being "light-years ahead" of their competitors. They come up with a product so revolutionary that it leaves the others in the dust. It's more than light-years ahead, it's a whole new product. How to describe it?

Sometimes they say it's a "quantum leap" ahead of the others. But how big a leap is that, really?

The nature of matter has been a mystery for thousands of years (and really, it still is). Contrary to our modern bias that ancient

people were not as smart as we are now, the ancient Greeks theorized about the existence of atoms. The thinker Democritus deduced that if you split a rock in half, then do it again, and again, and again, eventually you might come to a point where you simply cannot split it any more. That tiniest part he called an atom, meaning "indivisible."

This knowledge was interesting but of no fundamental meaning until thousands of years later. The advent of better technology let us investigate these tiny atoms. At first, it was thought that the atom looked like a solid little ball, but experiments soon showed that there were two separate parts—a nucleus in the middle made of particles called protons and neutrons, and an outer part containing particles called electrons. One model had the atom looking like a miniature solar system, with the nucleus acting like the Sun and the electrons orbiting like little planets.

This model sparked a flurry of science-fiction stories in which the solar system itself was just an atom in a greater universe of matter. This concept was really just a model, not designed to be a true picture of reality. Nevertheless, the idea still persists today in much of the public's mind.

However, the model turned out to be incorrect. At the very beginning of the twentieth century, a new physics was born. It was called **quantum mechanics,** and it postulated a horde of weird theories. One of them is that electrons are not free to orbit as they wish but instead are confined to specific distances from the nucleus. These distances are like steps in a staircase. You can be on the bottom step, or on the second or third step, but you can't be on the second-and-a-half step; there isn't any such place. If you are on the bottom step and try to get to the second, either you have enough energy to get there or you stay put.

So it goes for electrons. They stick to their specific orbit unless they get enough energy to jump to the next one. If even 99 percent of the energy needed to jump comes their way, they cannot do it. They need *exactly* the right amount to move to that next step, that next level. This jump became known as a **quantum leap.**

In reality, a quantum leap is a teeny-tiny jump. The distances are fantastically small, measured as billionths of a centimeter or less.

So you might conclude that an ad bragging about a product being a quantum leap over other products is silly, since it means it's ahead by only 0.00000000001 centimeters!

You might be surprised to find out that I have no problem with this phrase. I don't think it's bad at all! The actual distance jumped may be small, *but only on our scale*. To an electron it truly *is* a quantum leap, a sudden jump from one stage to the next. The phrase itself has nothing to do with the absolute distance the electron moved, but everything to do with its being a major leap forward, skipping the intervening space and landing in a new spot far ahead of where it was.

Sometimes people say that when something is easy, it isn't exactly rocket science. But in this case, maybe it is!

PART II

🐚 🐚 🐚 🐚 🐚 🐚

From the Earth to the Moon

The Earth is a big place. There are 511,209,977 square kilometers of it, give or take a kilometer or two, which might seem like room enough for everything. But even that much surface area isn't enough to contain all the bad astronomy out there. Not by a long shot. I wish it were at least true that it could be restricted to near-Earth space, but even then we run out of room pretty quickly. Still, there's a lot to be seen in our extended neighborhood. You need not even wait for nightfall. Most people might associate astronomy with nighttime, but we can scrounge up some during the day, too. As I write this the sky is a deep, rich blue, and the warm sunshine is blanketing my backyard. Just a few steps outside my house I can feel the warm embrace of an environment fraught with myths, misconceptions, judgment errors.

That cerulean-blue sky is a good place to start. True to the cliché, one day my five-year-old daughter asked me why the sky was blue, and I had to figure out how to answer her. I explained to her about molecules and sunlight, and the cosmic pachinko game played as the light from the sun makes its way to our eyes. When I was done, she thought about it for a second, and said, "All that stuff you just said doesn't make any sense."

I hope I've done better writing it all down in the next chapter.

But why stop with our air? We can move out of the atmosphere and peer back down on the Earth, seeing our frigid poles

37

and tropical equator. Why are those two locales different, and why does everything in between change from season to season? That's a fair question, too, and the cause is rooted in astronomy.

Moving a bit farther out, we encounter the Moon, our closest neighbor in the universe. I cannot think of any other object so loaded down with grossly inaccurate theories. The Moon only shows one face to us, but it does spin; it goes through phases that look like minature eclipses, but they are nothing of the sort; it looks unchanging and unchangeable, but that, too, is an illlusion. In the past, in the future, and even right now as you read these words, the Moon is being sculpted by unseen forces, just as it is profoundly changing the Earth. These same forces are at work throughout the universe, shaking mighty volcanoes, tearing apart stars, devouring entire galaxies.

If we can put a man on the Moon, you'd think we could stamp out most of the bad astronomy floating in the Earth's immediate vicinity.

4

❧❧❧❧❧❧

Blue Skies Smiling at Me:
Why the Sky Is Blue

In the course of every parent's life there comes the inevitable question from their child: "Why is the sky blue?" As we grow to adulthood we sometimes learn not to ask such questions, or we just forget how. The vast majority of adults in the world have seen a clear-blue sky tens of thousands of times, yet only a few know just *why* it's blue. If you don't know, don't fret: the question baffled scientists for hundreds of years. Nowadays we are pretty confident that we know the real reasons, but I've never heard of them being taught in schools. Even worse, a lot of web sites I've seen give an incorrect answer to the question. College textbooks on optics and atmospheric physics cover the topic correctly, but who wants those lying around the house?

Well, *I* do, but then I'm a huge geek. I'm operating on the principle that you are a normal human. And, lucky for you, the reason behind the blue sky isn't all that complicated, and it can be easily explained, even to a five-year-old. Let's start with some of the incorrect reasons given for the sky's cerulean hue.

Probably the most common idea is that the sky is blue because it reflects the blue color of the ocean. However, a moment's reflection (ha-ha) reveals that this can't be right: if it were true, the sky would look bluer when you are sailing on the ocean than when you are on land. But that's not the way it happens. It still looks just as blue from say, Kansas—a healthy hike from the nearest significant body of water—as it would from an ocean liner steaming its way from the United States to England.

Another commonly given incorrect answer is that blue light from the Sun scatters off dust in the air. As we'll see, this answer is close, and certainly better than the one about reflections off water, but dust is not the cause.

The correct answer, if you want details, is a little more involved. In the end we can simplify it for our hypothetical five-year-old, but first let's look at the whole problem.

When you examine most problems in astronomy, or for that matter in any other field of science, you'll commonly find that to get to the solution you need two separate lines of attack. The color of the sky is no exception. To understand the blueness we actually have to understand *three* things: just what sunlight is, how it travels through our atmosphere, and how our eyes work.

You may be surprised to learn that when it leaves the Sun's surface, sunlight is white. By this scientists mean it is actually a balanced combination of many colors. The individual colors like red, green, and blue are all produced by the complex physics near the sun's surface. The roiling, writhing gas making up the Sun's outermost layers produces light of all different colors. But when this light gets mixed together, it produces what looks to our eyes like white light. You can prove this for yourself: Hold a glass prism up to a beam of sunlight. When the sunbeam passes through the prism, the light gets "broken up" into its constituent colors. This pattern of colors is called a **spectrum.**

This same thing happens after a rainstorm. The raindrops suspended in the air act like little prisms, breaking up the white sunlight into a spectrum. That's how we get rainbows. The order of the colors in a rainbow is the same every time: red on the outside, then orange, yellow, green, blue, indigo, and finally violet, which makes up the innermost curve of the arc. This pattern may be tough to remember, so it's usually taught to students using the acronym ROY G BIV, like that's a common name or something. Still, that's how *I* remember it, so it must work.

Those colors are coming from the Sun all at the same time, but a funny thing happens on the way to the ground. Molecules of nitrogen and oxygen (N_2 and O_2) in the air can intercept that light. Almost like little billiard balls, **photons**—the fancy name for particles of light—bounce off these molecules and head off in a different direction every time they hit one. In other words, nitrogen and oxygen molecules *scatter* the incoming sunlight like bumpers in a pinball machine.

In the mid-1800s the brilliant British physicist Lord Rayleigh found out a curious thing: this scattering of light by molecules depends on the color of the light. In other words, a red photon is a lot less likely to scatter than a blue photon. If you track a red photon and a blue photon from the Sun as they pass through the air, the blue photon will bounce off its original course pretty quickly, while a red one can go merrily on its way all the way down to the ground. Since Lord Rayleigh discovered and quantified this effect, we call it Rayleigh scattering.

So, what does this have to do with the sky being blue? Let's pretend you are a nitrogen molecule floating off in the atmosphere somewhere. Nearby is another molecule just like you. Now let's say that a red photon from the Sun comes at you. As Lord Rayleigh found, you don't affect the red photon much. It pretty much ignores you and your friend and keeps heading straight down to the ground. In the case of this red light, the Sun is like a flashlight, a shining source of red light in one small part of the sky. All the red photons the Sun emits come straight from it to some observer on the ground.

Now let's imagine a blue photon coming in from the Sun. It smacks into your friend, rebounds off him, and obligingly happens to head toward you. From your point of view, *that photon comes from the direction of that molecule and not the Sun.* Your molecule friend saw it come from the direction of the Sun, but you didn't because it changed course after it hit him. Of course, after it hits you that photon can rebound off you and go off in another direction. A third nitrogen molecule would see that photon as coming from you, not the Sun or the first molecule.

Now you're a person again, standing on the ground. When a blue photon from the Sun gets scattered around, at some point it will hit some final air molecule near you, go through a final scattering, and head into your eye. To you that photon appears to

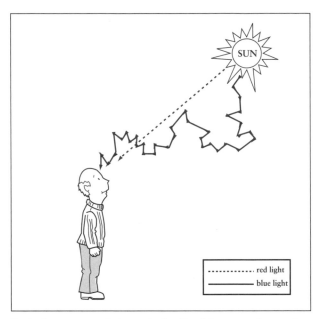

Red photons travel through the Earth's atmosphere relatively unimpeded, because of their relatively long wavelengths. Blue photons, however, with their considerably
shorter wavelengths, bump and careen around as they are
scattered by molecules in the air. By the time they reach
your eye, they appear to be coming from everywhere in
the sky, making it look blue.

come from that last molecule and not from the direction of the
Sun. These molecules are all over the sky, while the Sun is in one
little part of the sky. Since blue photons can come from any and all
of these molecules, the effect is that it looks like blue photons are
coming from every direction in the sky and not just the Sun.

That's why the sky looks blue. Those blue photons are converging down on you from all directions so that it looks to you
like the sky itself is giving off that blue light. The yellow, green,
orange, and red photons from the Sun get scattered much less than
do blue ones, and so they come straight at you from the Sun without having suffered all those scatterings.

At this point, you might reasonably ask why the sky isn't violet. After all, violet light is bent even more, and actually does scat-

ter more, than blue light. There are two reasons why the sky is blue and not violet. One is that the Sun doesn't give off nearly as much violet light as it does blue, so there's a natural drop-off at that color, making the sky more blue than violet. The other reason is that your eye is more sensitive to blue light than it is to violet. So, not only is there less violet light coming from the Sun but you're also less prone to notice it.

You can actually test this scattering idea for yourself in the safety of your own home. Get a glass of water and put a few drops of milk in it. Mix in the milk, then shine a bright white flashlight through the mixture. If you stand on the side of the glass opposite the flashlight, you'll see that the beam looks a bit redder. Go to the side and you will see the milk is bluer. Some of the blue photons from the flashlight are scattered away from the direction of the beam and go out through the sides of the glass, making the light look bluer. The light that passes all the way through is depleted in blue photons, so it looks redder.

This also explains the very common effect of red sunsets. One of the lesser known aspects of living on a big curved ball like the Earth is that as the Sun sets, the light travels through thicker and thicker air. The atmosphere follows the curve of the Earth's surface, so the light from an object that is straight overhead travels through far less air than the light from something near the horizon.

When the Sun is on the horizon, the sunlight travels through a lot more air than when it is up high during the day. That means there are more molecules, more scatterers, along its path, increasing the amount of scattering you'll see. Although blue light gets scattered a lot more than, say, yellow light, the yellow photons do scatter a little. When the Sun is on the horizon, the number of scatterers increases enough so that even green and yellow light can be pretty well bounced away into the rest of the sky by the time the sunlight reaches your eye. Since now the direct sunlight is robbed of blue, green, and yellow, only the red photons (which have longer wavelengths) make it through. That's why the Sun can be those magnificent orange or red colors when it sets, and also why the sky itself changes color near the horizon at the same time.

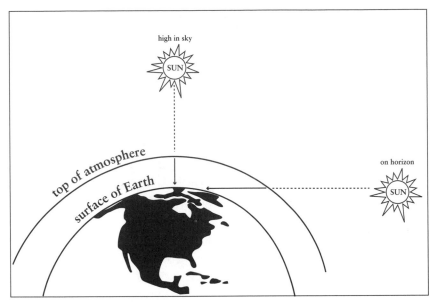

When you look straight up, you are looking through less air than when you look toward the horizon. Even green and yellow photons scatter away through the longer path they travel from the horizon, making the Sun look red or orange when it sets or rises.

It can look like that when it rises, too, but I think more people are awake at sunset than sunrise, so we see it more often in the evening. The Moon glows from reflected sunlight so it can change color, too, when it's on the horizon. Under unusually good conditions it can take on a startlingly eerie blood-red appearance.

This effect is amplified when there's more stuff in the air. Sometimes, when there are big volcanic eruptions, the sunsets are spectacular for quite some time afterwards. There's not much good to be said of explosive volcanic events, but they do put on quite an evening sky show for years.

There's another aspect of the curved atmosphere you've almost certainly seen as well. Have you ever noticed the Sun looking squashed when it sits on the horizon? The atmosphere, like a drop of water, can bend light. The amount that the light gets bent depends on the thickness of the air through which it travels. The more air, the more it's bent. When the Sun is on the horizon, the light from the bottom part of the Sun is traveling through more air than the top part. That bends the light more from the bottom part of the Sun. The air bends the light *up*, toward the top half, making the Sun look squashed. It doesn't get compressed left-to-right because the light from the left half of the Sun is moving through the same amount of air as the right half. As it sets the Sun looks normal horizontally, but it becomes more vertically challenged. The squashed, glowing, magenta Sun on a flat horizon is a sight not soon forgotten.

And now we have the three reasons the sky appears blue. First, the Sun sends out light of all colors. Second, the air scatters the blue and violet light from the Sun the most. And third, the Sun emits more blue than violet light, and our eyes are more sensitive to the blue light, anyway.

Now that we've established the color of the sky, we can tackle a related question that seems to cause a lot of anguish, and that is the color of the Sun.

If asked, I would say that the Sun is yellow. I think most people would, too. Yet we just went through a lot to show that sunlight is actually white. If the Sun is white, why do we think it looks yellow?

The key to the sentence above is the word "looks." Here's a sanity check: if the Sun were *really* yellow, clouds would look yellow, too. They reflect all the colors that hit them equally, so if they look white the Sun must be white. Don't believe me? Try this simple test: go outside and hold up a piece of white paper. What color is it? Okay, duh, it looks white. It looks white for the same reason clouds do. It reflects sunlight, which is white.

This brings us back to the original question: why does the Sun look yellow?

I have to cop out here. It's not really well-known why. Some people think the blue sky is to blame. If blue light is being scattered out of the direct sunlight hitting our eyes, the resulting color should look yellowish. While it's true that some blue light is scattered away, not *enough* of it is scattered to make the Sun very yellow. Even though a lot of blue photons are scattered away from the Sun to make the sky look blue, it's only a fraction of the *total* blue photons from the Sun. Most of them come straight to your eye, unimpeded by air molecules. So the relatively small number of photons making the sky blue doesn't really affect the intrinsic color of the Sun enough to notice.

Another common idea is that the Sun looks yellow because we are comparing it to the blue sky. Studies have shown that we perceive color not just because of the intrinsic properties of the light but also by comparing that color to some other color we see at the same time. In other words, a yellow light may look even yellower if seen against a background of blue. However, if this is why we see the Sun as yellow, clouds would look yellow, too, so this can't be right either.

There is another possibility. When the Sun is up high, you can never look directly at it. It's too bright. Your eyes automatically flinch and water up, making it hard to see straight. You can only see the Sun from the corner of your eye. Under those conditions it's not surprising that the colors may get a little distorted.

As was mentioned before, at sunrise and sunset the Sun can look remarkably red, orange, or yellow, depending on the amount of junk in the air. Also, the light is heavily filtered by the air, making the Sun look dim enough to be bearable to look at. So the only time of day we can clearly see the Sun is when it's low in the sky, which, not so coincidentally, is also when it looks yellowish or red. This may also play a part in the perceived color of the Sun. Since it looks yellowish at the only time we can really see it, we remember it that way. This is an interesting claim, although I have my doubts. I remember it most when the Sun is a glowing magenta or red ember on the horizon, and not yellow, so why don't I think the Sun is red?

I have heard some people claim the Sun *does* look white to them, but I wonder if they know that sunlight is supposed to be white, and have fooled themselves into thinking it *is* white to them. It still looks yellow to me, and I know better.

Clearly, there's more to the Sun than meets the eye.

So, after all this, I'll ask one more trick question: of all the colors of the rainbow, which color does the Sun produce the most? We know it produces less violet than blue; literally, fewer violet photons come from the Sun than blue. But which color is strongest?

The answer is: green. Surprise! So why doesn't the Sun *look* green? Because it isn't producing *only* green but a whole spectrum of colors. It just produces *more* green than any other color. When they are all combined, our eye still perceives the light as white.

Or yellow. Take your pick.

Okay, I lied a minute ago; I still have one more question. If the sky isn't blue because it reflects the color of the oceans, why *are* the oceans blue? Do they reflect the sky's color? No. Of course, they do reflect it a little; they look more steely on overcast days and bluer on sunny days. But the real reason is a bit subtler. It turns out that water can absorb red light very efficiently. When you shine a white light through deep water, all the red light gets sucked out by the water, letting only the bluer light through. When sunlight goes into water, some of it goes deep into the water and some of it reflects back to our eyes. That reflected light has the red absorbed out of it, making it look blue. So the sky is blue because it scatters blue light from the Sun, and the oceans look blue because that's the only light they let pass through.

❧❧❧

At the start of this section, I promised you'd understand all this well enough to explain it to a five-year-old. If a little kid ever asks you just why the sky is blue, you look him or her right in the eye and say, "It's because of quantum effects involving Rayleigh scattering combined with a lack of violet photon receptors in our retinae."

Okay, that might not work. In reality, explain to them that the light coming from the Sun is like stuff falling from a tree. Lighter things like leaves get blown all around and fall everywhere, while heavier things like nuts fall straight down without getting scattered around. Blue light is like the leaves and gets spread out all over the sky. Red light is like the heavier material, falling straight down from the Sun to our eyes.

Even if they still don't get it, that's okay. Tell them that once upon a time, not too long ago, nobody knew why the sky was blue. Some folks were brave enough to admit they didn't understand and went on to figure it out for themselves.

Never stop asking *why*! Great discoveries about the simplest things are often made that way.

5

❧ ❧ ❧ ❧ ❧ ❧

A Dash of Seasons:
Why Summer Turns to Fall

Some examples of bad astronomy are pernicious. They sound reasonable, and they even agree with some other preconceived notions and half-remembered high school science lessons. These ideas can really take root in your head and be very difficult to get out.

Perhaps the most tenacious of these is the reason why we have seasons.

Seasons are probably the most obvious astronomical influence on our lives. Over most of the planet it's substantially hotter in the summer than in the winter. Clearly, the most obvious explanation is our distance from the Sun. It's common sense that the closer you are to a heat source, the more heat you feel. It's also common sense that the Sun is the big daddy of all heat sources. Walking from underneath the shadow of a shade tree on a summer's day is all you need do to be convinced of that. It makes perfect sense that if somehow the Earth were to get closer to the Sun, it could heat up quite a bit, and if it were farther away our temperatures would dip. And hey, didn't you learn in your high school science class that the Earth orbits the Sun in an ellipse? So sometimes the Earth really *is* closer to the Sun, and sometimes it's farther away. This logic process seems to point inevitably to the cause of the seasons being the ellipticity of the Earth's orbit.

Unfortunately, that logic process is missing a few key steps.

It's true that the Earth orbits the Sun in an ellipse. We know it now through careful measurements of the sky, but it isn't all that obvious. For thousands of years it was thought that the *Sun* orbited the *Earth*. In the year 1530, the Polish astronomer Nicolaus Copernicus first published his idea that the Earth orbited the Sun. The problem is, he thought the Earth (and all the planets) moved in a perfectly circular path. When he tried to use that idea to predict the positions of the planets in the sky, things came out wrong. He had to really fudge his model to make it work, and it never really did do a good job predicting positions.

In the very early part of the 1600s, Johannes Kepler came along and figured out that planets move in ellipses, not circles. Here we are 400 years later, and we still use Kepler's discoveries to figure out where the planets are in the sky. We even use his findings to plan the path of space probes to those planets; imagine Kepler's reaction if he knew that! (He'd probably say: "Hey! I've been dead 350 years! What took you so long?")

But there's a downside to Kepler's elliptical orbits; they play with our common sense and allow us to jump to the wrong conclusions. We know that planets, including our own, orbit the Sun in these oval paths, so we know that sometimes we're closer to the Sun than at other times. We also know that distance plays a role in the heat we feel. We therefore come to the logical conclusion that the seasons are caused by our changing distance from the Sun.

However, we have another tool at our disposal beside common sense, and that's mathematics. Astronomers have actually *measured* the distance of the Earth to the Sun over the course of the year. The math needed to convert distance to temperature isn't all that hard, and it is commonly assigned as a homework problem to undergraduate-level astronomy majors. I'll spare you the details and simply give you the answer. Surprisingly, the change in distance over the course of the seasons amounts to only a 4-degree Celsius (roughly 7 degrees Fahrenheit) change in temperature. This may not surprise people from tropical locations, where the local temperature doesn't vary much over the year, but it may come as a shock to someone from, say, Maine, where the seasonal temperature change is more like 44 degrees Celsius (80 or so degrees Fahrenheit).

Clearly, something else must be going on to cause such a huge temperature variation. That something else is the tilt of the Earth's axis.

Imagine the Earth orbiting the Sun. It orbits in an ellipse, and that ellipse defines a plane. In other words, the Earth doesn't bob up and down as it orbits the Sun; it stays in a nice, flat orbit. Astronomers call this plane the **ecliptic**. As the Earth revolves around the Sun, it also spins on its axis like a top, rotating once each day. Your first impression might be to think of the Earth's axis pointing straight up and down relative to the ecliptic, but it doesn't. It's actually tilted by 23.5 degrees from vertical. Have you ever wondered why globe-makers always depict the Earth with the north pole pointing at an angle from straight up? Because it *is* tilted. It doesn't point up.

That tilt may not seem like a big deal, but it has profound implications. Here's an easy experiment for you: Take a flashlight and a piece of white paper. Darken the lights in a room and shine the flashlight straight down on the paper. You'll see a circle of bright light. Now tilt the paper so that the light shines down at about a 45-degree angle. See how the light spreads out? It's now an oval, not a circle. But more importantly, look at the brightness of the oval as you change the illumination angle. *It's dimmer.* The total light hitting the paper hasn't changed, but you've spread the light out by tilting the paper. More of the paper is lit, but each part of the paper has to share all the light, so there is less light for each part. If you tilt the paper more, the light gets even more spread out, and dimmer.

This is exactly what's happening to the Earth. Imagine for a moment that the Earth is not tilted, and that the axis really does point straight up and down relative to the ecliptic. Now pretend the Sun is a giant flashlight shining down on the Earth. Let's also say you are standing in Ecuador, on the Earth's equator. To you the Sun would be straight up at noon, with the sunlight hitting the ground straight on. The light is highly concentrated, just like it was when the paper was directly facing your flashlight in the experiment.

But now let's pretend you are in Minneapolis, Minnesota, which happens to be at 45 degrees latitude, halfway between the equator

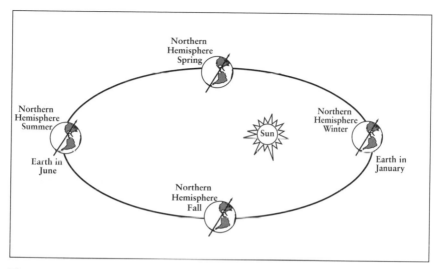

The seasons are caused by the Earth's tilt, and not because of its distance from the Sun. In the northern hemisphere, it's summer when the Earth's north pole points most toward the Sun, and winter when it points away. Note that the Earth is closest to the Sun during the northern hemisphere winter.

and the north pole. Again, sunlight is spread out like it was when you tilted the paper in the experiment. Since the Sun's light is what heats the Earth, there is less heat hitting the ground per square centimeter. The ground there doesn't get as much warmth from the Sun. The total light hitting the ground is the same, but it's spread out more.

To take matters to the extreme, imagine you're at the north pole. The sunlight there is hitting the ground almost parallel to it, and it gets spread out tremendously. Another way to think of this is that at the north pole, the Sun never gets very high off the horizon. This is like tilting your paper until the flashlight is shining almost along it. The light gets spread out so much that it barely does any good at all. That's why it's so cold at the north and south poles! The Sun is just as bright down there as it is in Ecuador and Minneapolis, but the light is spread out so much it can barely warm up the ground.

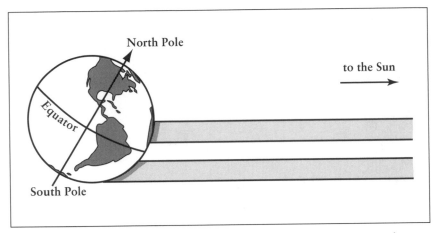

In the summer, the Sun is higher in the sky. Its light is more concentrated on the Earth's surface. In the winter, the Sun is lower, and the light gets spread out, heating the Earth less efficiently.

In reality, the Earth's axis is tilted, so matters are a bit more complicated. As the Earth orbits the Sun, the axis always points in the same part of the sky, sort of like the way a compass needle always points north no matter which way you face. You can imagine that the sky is really a crystal sphere surrounding the Earth. If you were to extend the axis of the Earth until it intersects that sphere, you'd see that the intersection doesn't move; to us on the surface of the Earth, it always appears to point to the same part of the sky. For those in the Earth's northern hemisphere, the axis points very close to the star Polaris. No matter what time of year, the axis always points in the same direction.

But as the Earth orbits the Sun, the direction to the *Sun* changes. Around June 21 each year the axis in the northern hemisphere is pointing as close as it can to the Sun. Six months later, it is pointing as far away as it can from the Sun. This means that for someone in the northern hemisphere the Sun is very high in the sky at noon on June 21, and very low in the sky at noon on December 21. On June 21, the sunlight is concentrated as much as it can be, and so it heats the ground efficiently. On December 21 the light gets spread out and it doesn't heat things up well. *That's* why it's hot in the summer and cold in the winter, and that's why we have seasons. It's not our *distance* from the Sun, but the *direction* to the Sun and therefore the angle of the sunlight that makes the difference.

Take a look at the diagram showing the Earth's axis relative to the Sun. Note that when the northern-hemisphere axis points toward the Sun, the southern-hemisphere axis points away, and vice versa. That's why people in the southern hemisphere celebrate Halloween in the spring and Christmas in the summer. I wonder if the song "I'm Dreaming of a Green Christmas" is popular in Australia . . .

There's an added tweak, too: because of our axial tilt the Sun gets higher in the sky in the summer, as we've seen. That means the path the Sun appears to travel in the sky is longer, so the Sun is up longer during the day. This, in turn, gives the Sun more time to heat up the Earth. Not only do we get more direct sunlight, the sunlight also lasts longer. Double whammy! In the winter the Sun doesn't get up as high, and so the days are shorter. The sun also has less time to heat up the ground, and it gets even colder. If the Earth were not tilted, days and nights would be 12 hours each, no matter where you were on the Earth, and we'd have no seasons at all.

Take another look at the figure on page 51. It shows that the Earth is actually closest to the Sun in January. This is the final nail in the coffin of the misconception that distance to the Sun is the main reason we have seasons. If that were true, we should have summer in January in the northern hemisphere and winter six months later in June. Since the opposite is true, distance must actually be a bit player in the seasons game.

However, it is not *completely* negligible. Distance does play a role in the seasons, although a minor one. For the folks in the northern hemisphere it means that winters should be a couple of degrees warmer on average than they would be if we orbited the Sun in a circle because we are closer to the Sun in the winter. Conversely, the summers are a couple of degrees cooler because we are farther away. It also means that people in the southern hemisphere should have hotter summers and colder winters than do those living in the northern hemisphere.

However, in reality, things are even more complicated. The southern hemisphere is mostly water. Check a globe and see for yourself if you like. Water is slower than land to heat up and cool off. This plays a role in the heat budget of the Earth, too. As it turns out, summers in the southern hemisphere are about as hot and winters are about as cold as they are in the northern hemisphere. The huge amount of water south of the equator acts as a kind of buffer, protecting that hemisphere from big temperature swings.

Amazingly, there is even more to this story. I said earlier that the Earth's axis is fixed in space, but I lied. Forgive me; I didn't want to make this too complicated at that point. The truth is, the Earth's axis does move, slowly, across the sky.

A slight digression: When I was a kid, my parents bought me a toy top. I used to love to spin it, watching it move across the floor in funny patterns. I also noticed that as it began to slow its spin, it would start to wobble. I was too young to understand it then, but I now know that the wobble is due to the interplay of complicated forces on the spinning top. If the axis of the top is not exactly vertical, gravity pulls the top off-center. This is called a *torque*. Because the top is spinning, you can think of that force being deflected horizontally, making the top slowly wobble. The same thing would happen if the top were spinning in space and you poked it slightly off center. The axis would wobble, making little circles; the bigger the poke, the bigger the circle it would make.

This wobble is called **precession,** and it is caused by any tug on the top that is not lined up with the axis. It happens for any spinning object that experiences some kind of force on it. Of course, the Earth spins, too, just like a top, and there does happen to be a force on it: the Moon's gravity.

The Moon orbits the Earth and pulls on it with its gravity. The Moon's tug on the Earth acts like an off-axis poking, and, sure enough, the Earth's axis precesses. It makes a circle in the sky that is 47 degrees across, exactly twice the size of the Earth's axial tilt, and that's no coincidence. The amount of the Earth's tilt with respect to the ecliptic, the orbital plane, doesn't change; it's always

23.5 degrees. However, it's the *direction* in the sky that changes with time.

The effect is slow; it takes about 26,000 years for the Earth's axis to make a single circle. Still, it's measurable. Right now the Earth's northern axis points toward Polaris, the Pole Star, which is how the star got that name in the first place. But it wasn't always pointed that way, nor will it be. As the axis precesses, it points to a different part of the sky. Back in 2600 B.C. or so it pointed toward Thuban, the brightest star in the constellation Draco. In A.D. 14,000 or so it will point near the bright star Vega.

For astronomers, precession is a bit of a headache. To measure positions of astronomical objects, astronomers have mapped out the sky in a grid much like the way cartographers have mapped the surface of the Earth into latitude and longitude. The north and south poles on the sky correspond to those same poles on the Earth, but the sky's north pole moves due to precession. Imagine trying to figure out directions on the Earth using north, south, east, and west if the north pole kept wandering around. You'd need to know just where the north pole was to know in which direction you needed to go.

Astronomers have the same problem on the sky. They must account for the precession of the Earth's axis when they measure an object's position. The change is small enough that most sky maps need to be updated only every 25 to 50 years. This is particularly important for telescopes like the Hubble Space Telescope, which must point with incredible accuracy. If the precession is not included in the calculation of an object's position, the object might not even be in the telescope's field of view.

Precession has an immediate impact on astronomers, but a much slower one on the seasons. Right now, the Earth's north axis points toward the Sun when the Earth is farthest from the Sun. But due to precession, 13,000 years from now—half a precession cycle—the Earth's north pole will be pointed *away* from the Sun at the farthest point in its orbit, and *toward* the Sun six months later, when the Earth is closest to the Sun. Seasons will be reversed relative to the orbit now.

So, half a precession cycle from now, the northern hemisphere of the Earth will experience summer when the Earth is closest to the Sun, amplifying the heat. It'll also be winter when we're farther from the Sun, amplifying the cold. Seasons will be more severe. In the southern hemisphere, the seasons will be even milder than they are now, since they'll have summer when we are farther from the Sun and winter when we are closer.

This works the other way, too: 13,000 years in the past, the seasons were reversed. Summers were hotter and winters were colder in the northern hemisphere. Climatologists have used that fact to show that things might have been profoundly different back then. The slow change in the direction of the Earth's axis might have even been the cause of the Sahara becoming a desert! On a year-by-year basis precession is barely noticeable, but over centuries and millennia even small changes add up. Nature is usually brutal and swift, but it can also display remarkable subtlety. It just depends on your slant.

6

Phase the Nation:
The Moon's Changing Face

I never know whether to be surprised at the fact that, of the all the topics touched by bad astronomy, the Moon has the longest tally.

I'm surprised because the Moon is probably the most obvious of all astronomical objects. Some might argue the Sun is, but you can never really look right at the Sun. It's always in the corner of your eye but never fully in it.

The Moon is a different story. When the night is dark, and even the crickets have gone to sleep, the full Moon shines down in blazing contrast to the black sky. Even as the thinnest of crescents the Moon commands attention, hanging low in the west after sunset. Whether high in the sky or low near the horizon, it dominates the night.

So it surprises me that there is so much misunderstanding about the Moon. I would think that since it's such a common sight, it would be the best understood.

But perhaps that's naive. After all, the more we know about something, the more room there is to *mis*understand it. So it is with the Moon.

Why does the Moon look bigger near the horizon than when it's overhead? Why does it have phases? How does it cause tides? How can it be up during the day? Why does it show only one face to the Earth? Which part is the dark side?

These topics all have some pretty hefty bad astronomy associated with them, and I promise we'll get to all of them. But first things first. The most obvious aspect of the Moon is that it *changes*. Even the least attentive of sky watchers will notice that sometimes the Moon is a thin crescent and sometimes it's a big fat white disk hanging in the sky. In between those times, it can be half full or partially full. Sometimes it's gone altogether! These shape changes are called the **phases** of the Moon. What causes them?

A lot of people think it's due to the shadow of the Earth falling on the Moon. The Moon is a big sphere, so when it's almost all the way in the Earth's shadow, the thought goes, the Moon is a crescent. When it's fully out of the shadow, it's full.

That's a clever idea, but incorrect. The Sun is the major source of light in the solar system. That means the Earth's shadow always points away from the Sun. That, in turn, means the Moon can only be in the Earth's shadow when it's on the opposite side of the sky from the Sun. But the Moon can't always be in Earth's shadow, especially when it's near the Sun in the sky. We also know that when the Moon gets directly between the Earth and the Sun we get a total solar eclipse. That's a pretty rare event, yet the Moon's phase changes every night. Clearly, the Earth-shadow theory cannot be correct, and something else must be going on.

So what do we know about the Moon? Well, it's a big ball, and it orbits the Earth once a month. Actually, the word "month" is derived from the same root as the word "Moon." The phases change as the Moon goes around us, which is a clear indication that they must have something to do with the orbit. In science, it's usually best to take stock with *what* you see before trying to figure out *why* you're seeing it. So, let's take a look at the phases and start at the start.

New Moon marks the beginning of the lunar cycle of phases, which is why it's called new. When the Moon is new, it's completely dark. This happens when it's near the Sun in the sky. Since the Sun is so bright and the Moon is dark, the new Moon can be very difficult to see. The Islamic month, for example, begins at the time the very earliest new Moon can be spotted, and so the followers of Islam keep very careful records and have keen-eyed observers ready to see it as early as possible.

First quarter is when the Moon is half lit, confusingly enough. It's called first quarter because the Moon is lit like this when it's one-quarter of the way around the Earth from the Sun, roughly one week after new Moon. For people in the northern hemisphere of the Earth, this means the right-hand side of the Moon—the side facing the Sun—is lit and the left-hand side is dark. For people in the southern hemisphere the reverse is true, since, to the view of people in the north, people in the south are upside-down.

A week later, the Moon is **full.** The whole disk is evenly illuminated. When the Moon is full it's opposite the Sun in the sky, and it rises when the Sun sets.

A week after that, the Moon is at **third quarter.** Just like when it's at first quarter, the Moon is half lit, and, also like first quarter, the half facing the Sun is lit. This time, though, it's the other half that's lit. From the northern hemisphere, the *left* half is lit and the *right* half is dark. Reverse that if you're south of the equator.

Finally, a week later, the Moon is new again, and the cycle repeats. There are also names for the phases of the Moon when it's between these four major ones. As more of the visible part of the Moon becomes lit, we say it is **waxing.** When the Moon is between new and first quarter, it's still crescent shaped but it's getting fatter, approaching half full. We say the Moon is now a **waxing crescent.** After it's half full and approaching full, it's in the **gibbous phase,** or, more accurately, **waxing gibbous.** After it's full, it starts getting smaller. This is called **waning.** The Moon is **waning gibbous** from full to third quarter, and a **waning crescent** from third quarter to new.

So now we have names for all those shapes. The question remains, why does the Moon go through phases? Now that we've looked at them, we're closer to figuring that out. However, there's one more thing I want you to do. Go get a ping-pong ball or a baseball. Don't have one? That's okay, you can use your imagination.

Imagine that you are holding a white styrofoam ball. This is our model of the Moon. You will be the Earth and, for this demonstration, a lamp across the room will be the Sun. Before we start the demo, let's think about this for a second: when you hold up the ball, half of it will be lit by the lamp and half will be in shadow. That seems obvious, but it's crucial to understanding phases. No matter how you hold the ball, half will always be lit, and half dark. Got it? Okay, let's set the Moon in motion.

Let's start at new Moon. When it's new, the Moon is between the Sun and Earth. Imagine holding the Moon up so that it lines up with the Sun. From your point of view, the Sun is glowing brightly, but the Moon itself is dark. That's because the side of the Moon being lit by the Sun is facing *away* from the Earth. From the Earth, we only see the side that is not lit by the Sun, so it's dark.

Now move the Moon one-quarter of an orbit around from the Sun. The Sun is off to the right, and so the right-hand side of the Moon is lit. The left-hand side is dark. Remember, half the Moon is always lit by the Sun, but when it's in this part of the orbit, we only see half of that half. We see one quarter lit up.

Now turn so that the Moon is opposite the Sun. With your back to the Sun, you see the entire half of the Moon facing you lit up, and it's full. (Incidentally, that's why photographers like to take portrait shots with the Sun over their shoulder: that way, your face is fully illuminated by the Sun and there are no shadows on it. Of course, you have to squint because the Sun's in your eyes, but that's a sacrifice you make for a good shot.)

Finally, turn so that the Moon is three quarters of the way around in its orbit. The Sun is now off to the left, and the left-hand side of the Moon is lit. Again, of course, really half the Moon is lit, but you see only half of that half. This time, since the Sun is to the left, you see the left half lit up. The right side is in shadow, and it's dark.

That's what causes the phases. It's not the Earth's shadow at all. The Moon has phases because it's a ball, with one half lit by the Sun. Over a month, its position relative to the Sun changes, showing us different parts of it being lit up.

Once you understand this, an interesting side effect can also be seen. For example, at new Moon, the Moon always appears near the Sun in the sky. That means it rises at sunrise and sets at sunset. When the Moon is full, it's opposite the Sun in the sky. It rises at sunset and sets at sunrise. The Moon is like a giant clock in the

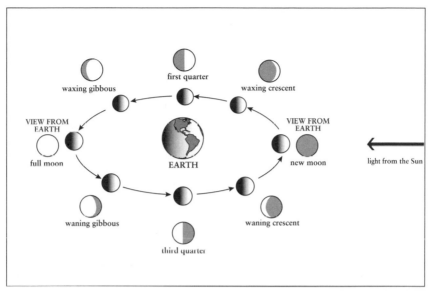

The Moon's phases are an effect of geometry, and not due to the Earth's shadow on the Moon. In this diagram, the Sun is off to the right. The position of the Moon is shown in the inner circle, while the phase seen by someone on the Earth is shown in the outer circle. The Moon is new when it is closest to the Sun in the sky, and full when it's farthest from the Sun. The other phases happen as the Moon orbits the Earth.

sky. If the full Moon is high in the sky, it must be near midnight (halfway between sunset and sunrise). If it's getting low in the west, sunrise cannot be far behind.

You can get even fancier with the quarter-Moon phases. The first-quarter Moon is one-quarter of the way around from the Sun, and is high in the sky at sunset (90 degrees away from the Sun). So it rises at noon, and sets at midnight. It's another common misconception that the Moon is only up in the sky at night. When it's at first quarter it can be seen easily in the afternoon sky; the third-quarter Moon can also be seen in the sky after sunrise, since it sets at noon.

Another obvious feature of the Moon is that its brightness changes with phase. This seems pretty obvious; after all, there is more of it lit up when it's full than when it's half full. You might think that it is twice as bright then.

That turns out not to be the case. Like everything else in astronomy, there's more to this story. Careful measurements of the Moon's brightness show that it can be up to *ten times* brighter when it's full compared to its first quarter.

There are two reasons for this. One is that when the Moon is full, the Sun is shining straight down on it from our viewpoint. When the Sun is directly overhead here on Earth, there are no shadows, and when it's low in the sky shadows are long. The same is true for the Moon. There are no shadows on the surface when the Moon is full. When it's at first quarter there are lots of shadows, which darken the surface, making the Moon look less bright overall. When the Moon is full, those shadows aren't there, and so it has more than twice as much lit surface from our view than when it's at first quarter.

The other reason has to do with the Moon's surface. Meteorite impacts, ultraviolet radiation from the Sun, and the violent temperature changes from day to night on the Moon have eroded the top centimeter or so of the lunar surface. The resulting powder is extremely fine, like well-ground flour. This powder has a peculiar property: it tends to reflect light directly back to the source. Most objects scatter light every which way, but this weird soil on the Moon focuses much of the light back toward the source. This effect is called **back-scatter.**

When the Moon is half full, the Sun is off to the side as seen by us. That means the lunar soil tends to reflect that light back toward the Sun, away from us. When the Moon is full, the Sun is directly behind us. Sunlight that hits the Moon gets reflected, preferentially back to the Sun, but we are in that same direction. It's as if the Moon is focusing light in our direction. This effect, together with the lack of shadows, makes the full Moon much brighter than you might expect.

Even the new Moon can be brighter than you expect. Normally, the new Moon is dark and difficult to spot. But sometimes, just after sunset, you can see the crescent Moon low in the sky. If you look carefully, sometimes you can see what looks like the outline of the rest of the Moon, even though it's dark.

Your eyes aren't playing tricks on you. This effect is called **earthshine.** From the Moon, the Earth goes through phases, too. They are opposite the Moon's phases, so when the Moon is full as seen from the Earth, the Earth would be new as seen from the Moon, and so on. The Earth is physically bigger than the Moon, and it also reflects light more efficiently. The full Earth as seen from the Moon would look many times brighter to you than the full Moon does on the Earth.

This brightly lit Earth illuminates the new Moon pretty well, faintly lighting what would normally be the dark part of the Moon's surface. If you look through a telescope or a pair of binoculars, there's even enough light to spot craters on the surface. The effect is even more amplified if the lit side of the Earth is covered by clouds, making the Earth an even better reflector of sunlight.

Earthshine is a pretty name for this, but there's an even more poetic one: it's called "The old Moon in the new Moon's arms."

The phases of the Moon are both more complicated and more subtle than you might have thought. If you had any misconceptions about them before reading this section, let's hope it was just a phase.

7

⤪ ⤪ ⤪ ⤪ ⤪ ⤪

The Gravity of the Situation: The Moon and the Tides

"There is a tide in the affairs of men . . . "
                              —*Julius Caesar* by William Shakespeare

If I had a nickel for every time I am asked about tides . . . I'd have a *lot* of nickels.

There are a lot of misconceptions about tides. Anyone who has spent a day at the beach knows about tides; the difference between high and low tide can be substantial. But the details of tides can be a bit weird. For example, there are roughly two high tides and two low tides a day. I get questions about this all the time. Most people have heard that the Moon's gravity causes tides, so why are there two high tides each day? Shouldn't there be only one high tide, when the Moon is overhead, with a low tide on the opposite side of the Earth?

When I wrote a web page about tides, and again while researching them for this chapter, I couldn't find a single source that made any sense. Different web pages and different books all had different explanations. Some made sense for a while, then said something clearly wrong. Others started off wrong and got worse from there. Most are close, showing that the explanation relies on several different factors. What's worse: I wrote a draft of this chapter and even had it submitted for the book, then realized what I said was essentially wrong! What you'll read here is now correct. It's funny, too—even people who get tides right rarely take the discussion far enough. Tides have far-reaching consequences, from locking together the Moon's spin and orbital motion to the volcanoes on Jupiter's moon Io. Tidal forces can even cause entire galaxies to be ripped apart, torn to shreds by even bigger galaxies.

When astronomers talk about tides, we usually don't mean the actual movement of water. We are using the term as a shorthand for the **tidal force**. This is a force much like gravity, and in fact is related to gravity. We're all aware of gravity from the first time we try to stand up and walk. As we age, we become increasingly aware of it. For me, it seems harder to get out of bed every day, and easier to drop things. Sometimes I wonder if the Earth is pulling harder on me each day.

It doesn't really, of course. Gravity doesn't change with time. The *force* of gravity, the *amount* that it pulls on an object, depends on only two things: the mass of the object doing the pulling, and how far away it is.

Anything with mass has gravity. You do, I do, planets do, a feather does. I can exact a minute amount of revenge on Earth's gravity knowing that I am pulling back on the Earth as well. The amount I am pulling is pretty small, sure, but it's there. The more massive the object, the more it pulls. The Earth has a lot more mass than I do (something like 78,000,000,000,000,000,000,000,000 times as much, but who's counting?), so it pulls on me a lot harder than I do on it.

If I were to get farther away from the Earth, that force would weaken. As a matter of fact, the force drops with the square of my distance; that is, if I double my distance, the force drops by a factor of $2 \times 2 = 4$. If I triple my distance, it drops by $3 \times 3 = 9$, and so on.

That does not mean that I feel one-quarter of the gravity if I climb a ladder to twice my height, though! We don't measure distance from the surface of the Earth, we measure it from its center. A few hundred years ago, Sir Isaac Newton, the seventeenth-century philosopher-scientist, showed mathematically that as far as distance is concerned, you can imagine that all the mass of the Earth is condensed into a tiny point at its center, so it's from there that we measure distance.

The Earth's radius is about 6,400 kilometers (4,000 miles), so for me to double my distance, I'd have to book a flight on a rocket: I'd need to get an additional 6,400 kilometers off the ground, nearly one-sixtieth of the way to the Moon. Only there would I feel like I weigh a quarter of what I do now. It seems like a rather drastic way to lose weight.

Because the Moon is smaller and less massive than the Earth, you would feel a gravity about one-sixth that of the Earth's if you were standing on the lunar surface. That's still a substantial pull. Of course, the Moon is pretty far away, so its gravitational effect here on Earth is much smaller. It orbits the Earth at an average distance of about 384,000 kilometers (240,000 miles). From that distance its gravity drops by a factor of nearly 50,000, so we can't feel it.

But it's there. Gravity never goes away completely. Although on the Earth the force of gravity from the Moon is terribly weak, it still extends its invisible hand, grasping our planet, pulling on it.

That grasp weakens with distance, giving rise to an interesting effect on the Earth. The part of Earth *nearest* the Moon feels a stronger pull than the part of the Earth *farthest* from the Moon. The difference in distance—the diameter of the Earth—means a difference in gravity. The near side of the Earth feels a pull about 6 percent stronger than the far side. This difference in pull tends to stretch the Earth a little bit. It's because the *gravity* is *different* from one side of the Earth to the other, so we call it **differential gravity.**

Gravity always attracts, so the force of lunar gravity is always a pull *toward* the Moon. So, you would think, since the near side of the Earth feels a stronger pull, water would pile up there, giving us a high tide. On the far side of the Earth there should be a low tide, a flattening, perhaps, because even though the force is weaker, it still points toward the Moon.

But we know that's not right. There are *two* high and *two* low tides a day. That means at any one time there must be a high tide on the opposite side of the Earth from the Moon as well. How can this be?

Clearly, differential gravity isn't enough to explain tides. For the complete picture, we have to look once again to the Moon.

Allow me to digress for a moment.

A couple of years ago, my two good friends Ben and Nicky got married. They asked my then-three-year-old daughter Zoe to be the flower girl. The ceremony was lovely, and afterwards at the reception we all danced. Zoe wanted to dance with me, and what proud father could say no?

So I took her hands and we danced in a circle. I had to lean backwards a little to make sure we didn't topple over, and as I swung her around I couldn't help noticing that the circle she made on the floor was big, and the one I made was small. Since my mass was about five times what hers was, she made a circle five times bigger than mine.

So what does this have to do with tides? *Everything*. Our little dance is a tiny version of the same tango in which the Earth and Moon participate. Instead of holding each other's hands, the Earth and Moon use gravity to embrace. And just like Zoe and me, they both make circles.

Since the Moon's mass is one-eightieth the mass of the Earth, the effect of the Moon's pull on the Earth is one-eightieth the effect of the Earth's pull on the moon. Like my daughter making a bigger circle on the dance floor than I did, the Moon makes a big circle around the Earth, but the Earth also makes a little circle at the same time.

This means that the Moon and the Earth are actually orbiting a point in between the two bodies, as if all the mass in the Earth-Moon system is concentrated there. This point is called the **center of mass,** or technically the **barycenter.** Since the Earth is about 80 times the mass of the Moon, the center of mass of the whole system is about one-eightieth of the way from the center of the Earth to the center of the Moon. That's about 4,800 kilometers (3,000 miles) or so from the center of the Earth, or about 1,600 kilometers (1,000 miles) beneath the Earth's surface. If you could watch the Earth from outer space, you'd see it make a little circle centered on a point 1,600 kilometers beneath its surface, once every month. In a very real sense, the center of mass of the Earth (which is basically the center of the Earth itself) is orbiting the center of mass of the Earth-Moon system, making that little circle once a month.

This has some interesting implications. To see this, think about the astronauts on board the space station. They float freely, as if there is no gravity. In fact, they feel gravity almost as strongly as we do here on the surface of the Earth; after all, they are only a few hundred kilometers high, not much compared to the 6,400-kilometer radius of the Earth. The astronauts float because they are in free fall; the Earth is pulling them down, so they fall. But they have so much sideways velocity that they basically keep missing the Earth. Their orbit carries them along a curve that has the same curvature of the Earth, so they continuously fall but never get any closer to the ground.

An astronaut standing on a scale in the space station would measure her weight as zero because she is falling around the center of the Earth. Gravity affects her, but she cannot feel it. This is always true for an orbiting object.

But remember, the center of the Earth is orbiting the Earth-Moon barycenter, too. So even though the center of the Earth is *affected* by gravity from the Moon, someone standing there would not actually *feel* that force. They would be in free fall!

But someone standing under the Moon on the Earth's surface *would* feel the Moon's pull. Someone standing on the opposite side would, too, but more weakly. But since the force felt from the Moon's gravity is zero at the Earth's center, we can measure the Moon's gravity *relative to the center of the Earth*. To someone on the side of the Earth nearest the Moon, there would be a force felt toward the Moon. Someone in the center of the Earth feels no force (remember, they are in free fall). But the person on the far side of the Earth feels less force toward the Moon than the person at the center of the Earth. But what's smaller than zero force? A negative force; in other words, a positive force in the other direction, *away from the Moon*.

It seems paradoxical that gravity can act in such a way as to make something feel a force away from an object, but in this case

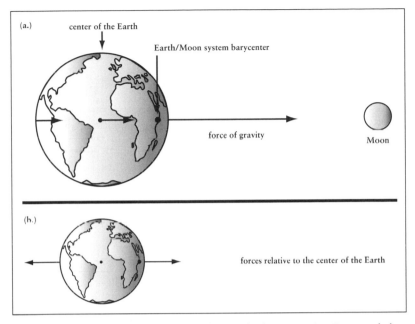

The force of the Moon's gravity on the Earth always "points" toward the Moon. The force gets weaker with distance so that the near side of the Earth is pulled toward the Moon more strongly than the far side. When the force of the Moon's gravity is calculated relative to the Earth/Moon center of mass, the far side of the Earth actually feels a force pointing *away* from the Moon, while on the near side the force is still toward the Moon. This results in a stretching of the Earth, which is why we have two high tides a day.

it's because we are measuring that force relative to the center of the Earth. When you do that, then you do indeed get a force pointing away from the Moon on the far side of the Earth.

That is why we have *two* high tides. There is a net force toward the Moon on the near side, and a net force away from the Moon on the far side. The water follows those forces, piling up in a high tide on opposite sides of the Earth. In between the two high tides are the low tides, and of course there are two of them as well. As a point on the Earth rotates under the high tide bulge, the water rises. A few hours later, when the Earth has rotated one-quarter of the way around, that point is now under a low tide, and the water has receded. One-quarter of the way around again and you've got a high tide. On and on it goes, with high and low tides alternating roughly every six hours.

But not *exactly* six hours. If we could hold the Moon stationary for a while, you would indeed feel two high (and two low) tides a day, separated by 12 hours. But as we saw in the last chapter, the Moon rises about an hour later every day, because as the Earth spins, the Moon is also orbiting the Earth. The Moon moves during that day, so we have to spin a little bit extra every day to catch up to it. So instead of there being 24 hours between successive moonrises, there are actually about 25. That means there is a little extra time between high tides; half of that 25 hours, or 12.5 hours. The time of high and low tides changes every day by about a half hour.

An aside: Most people think that only water responds to these tidal forces. That's not true; the ground does, too. The solid Earth isn't really all that solid. It can bend and flex (ask anyone who's ever been in an earthquake). The forces from the Moon actually move the Earth, shifting the ground up and down about 30 centimeters (12 inches) every day. You can't feel it because it happens slowly, but it does happen. There are even atmospheric tides. Air flows better than water, resulting in even more movement. So, the next time someone asks you if the Earth moved, say yes, about a third of a meter.

Incidentally, this puts to rest a common misconception about tides. Some people think that tides affect humans directly. The idea I usually hear is that humans are mostly water, and water responds to the tidal force. But we can see that idea is a bit silly. For one thing, air and solid ground respond to tides as well. But more importantly, humans are too small to be affected noticeably by the tides. The Earth has tides because it's big, thousands of kilometers across. This gives the gravity from the Moon room to weaken. Even a person two meters tall (6 feet 6 inches) feels a maximum difference in gravity of only about 0.000004% from head to foot. The tidal force across the Earth is over a million times stronger than that, so needless to say the tidal force across a human is *way* too small to be measured. Actually, it's completely overwhelmed by the natural compression of the human body in the standing position; you shrink from gravity more than you are stretched by tides. Even large lakes can barely feel tides; the Great Lakes, for example, have a change in height of only four or five centimeters due to tides. Smaller lakes would have an even smaller change.

As complicated as all that sounds, amazingly, we aren't done yet. Tides due to the Moon are only half the issue. Well, actually, they're two-thirds of the issue. The other third comes from the Sun.

The Sun is vastly more massive than the Moon, so its gravity is far stronger. However, the Sun is a lot farther away. The Earth orbits the Sun in the same way the Moon orbits the Earth, so the same idea applies. The Earth feels a gravitational pull toward the Sun and a centrifugal force away from it. If you do the math, you find out that tides due to the Sun are roughly half the strength of the lunar tides. In the tidal game mass is important but distance even more so. The nearby, low-mass Moon produces more tidal force on the Earth than the much more massive but much farther away Sun. Of the total tidal force exerted on the Earth, two-thirds is from the Moon and one-third is from the Sun.

The Earth is in a constant, complicated tug of war between the Sun and the Moon. There are times when the two objects' forces are in a line. As we saw in the last chapter, "Phase the Nation," when the Moon is new it is near the Sun in the sky, and when it's full it's opposite the Sun. In either case the tidal forces from the Moon and the Sun line up (because, remember, high tides occur simultaneously on opposite sides of the Earth, so it doesn't really matter which side of the Earth you are on), and we get extra-high high tides. It also means the low tides line up, so we get extra-low low tides. These are called **spring tides**.

When the Sun and Moon are 90 degrees apart in the sky, their forces cancel each other out a bit, and we get tides that aren't quite as low or as high (it's like a lower high tide and higher low tide). These are called **neap tides**.

Even worse, the Moon orbits the Earth in an ellipse, so sometimes it's closer to us than other times, and the forces are that much greater. The *Earth* orbits the *Sun* in an ellipse, too, so we get more exaggerated tides during the time of closest approach to the Sun as well (around January 4 each year). If these two events—closest

Moon, and closest point to the Sun—happen at the same time, we get the biggest possible tides. It's not really as big an effect as all that, though; it's only a few percent more. But as you can see, tides are complicated, and the force is never constant.

But there's no reason to stop here. There is another effect. It's subtle, but the implications are quite profound.

As I mentioned, the Earth is spinning on its own axis while the Moon orbits us. The water responds quickly to the tidal force, and "piles up" under the Moon and on the side of the Earth opposite the Moon. However, the Earth is spinning, and its spin is faster (one spin a day) than the Moon's motion around the Earth (one orbit a month). The water wants to pile up under the Moon, but friction with the spinning Earth actually sweeps it forward a bit, ahead of the Moon. The tidal bulge, as it is called, does not point directly to the Moon, but a little in front of it.

So picture this: the bulge nearest the Moon is actually a bit ahead of the Earth-Moon line. That bulge has mass—not a lot, but some. Since it has mass, it has gravity, and that pulls on the Moon. It pulls the Moon *forward* a bit in its orbit. It acts like a small rocket, pushing the Moon ahead a little. When you push an orbiting object forward, it goes into a higher orbit, that is, one with a larger radius. So, as the tidal bulge on the Earth pulls the Moon forward, the Moon gets farther away from the Earth. This effect has been measured quite accurately. The Moon is actually farther away now than it was a year ago by about 4 centimeters (1.5 inches). Next year it'll be another 4 centimeters farther away, and so on.

Of course, the Moon is pulling on that tidal bulge as well. If the bulge is *ahead* of the Moon, then the Moon is *behind* the bulge (relative to the rotation of the Earth). That means it's pulling the bulge backwards, slowing it down. Because of friction with the rest of the Earth, this slowing of the bulge is actually slowing the rotation of the Earth! This is making the day get longer. Again, the effect is small but measurable.

Besides the phase, the most obvious feature of the Moon is that it always shows the same face to us (described in chapter 3, "Idiom's Delight"). This is because the Moon spins once on its axis in the same amount of time it takes to orbit the Earth once.

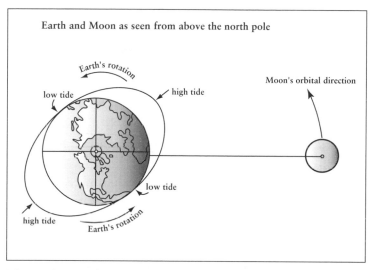

The Earth spins faster (once a day) than the Moon moves around the planet (once a month). A bulge caused by the lunar tide is swept ahead of the Moon by the Earth's rotation. This in turn tugs on the Moon, pulling it faster in its orbit, and moving it away from the Earth by about 4 centimeters per year. It is also slowing the Earth's spin at the same time.

This timing may seem like a miraculous coincidence, but it isn't. Tides force this situation.

All the time the Moon's gravity is exerting a tide on the Earth, the Earth is doing the same thing to the Moon. But the tides on the Moon are 80 times the force of the ones on the Earth, because the Earth is 80 times more massive than the Moon. All of the tidal effects on the Earth are also happening on the Moon, but even faster and stronger.

The Earth raises a big tide on the Moon, stretching it out. There are two high tide bulges on the Moon, right in its solid rock. When the Moon formed, it was closer to the Earth and rotated much faster. The tidal bulge raised by the Earth on the Moon started to slow the Moon's rotation, just as the Earth's high-tide bulge does here. As the Moon slipped farther from the Earth, its rotation slowed, until the rotation period was the same as its revolution period (in other words, its day equaled a month). When that happened, its bulges lined up with the Earth, and the rotation of the Moon became constant; it stopped slowing.

That's why the Moon always shows one face. It's rotating, but the tidal force, well, *forced* this to happen. It's not coincidence, it's science!

Remember, too, the Earth's rotation is slowing down. Just as the Moon did eons ago, eventually the Earth's rotation will slow down so much that the tidal bulge on the Earth will line up exactly between the centers of the Earth and the Moon. When this happens, the Moon will no longer be pulling the bulge back, and the Earth's spin will stop slowing. The Earth's day will be a month long (and by then the Moon's recession will mean that the month will be longer too, about 40 days). In this time, far in the future, if you were to stand on the Moon and look at the Earth, you would always see the same face of the Earth, just as we see one face of the Moon from the Earth.

This kind of change due to tides is called **tidal evolution,** and it has affected the Earth and the Moon profoundly. When they were young, the Earth and the Moon were closer together and they both spun much more quickly. But over the billions of intervening years, things have changed drastically.

Once the Earth is rotationally locked with the Moon, there will be no more evolution of the Earth/Moon system from mutual tides. However, there will still be tides from the Sun. They would affect the system, too, but by the time all this happens, the Sun will be well on its way to turning into a red giant, frying the Earth and the Moon. We'll have bigger problems than tides on our hands at that point.

Of course, we aren't the only planet with a moon. Jupiter, for example, has dozens. The tides that Jupiter raises on its moons are hellish; the planet is over 300 times the mass of the Earth. Little Io, a moon of Jupiter, orbits its planet at the same distance the Moon orbits the Earth, so it feels tides 300 times stronger than does our Moon. Io is also tidally locked to Jupiter, so it spins once an orbit. If you could stand on Jupiter, you'd always see the same face of Io.

But Jupiter has lots of moons, and some of them are big. Ganymede, for example, is bigger than the planet Mercury! These moons all affect each other tidally, too. When one moon passes another, the differential gravity squeezes and stretches the moons, flexing them.

Have you ever taken a metal coat hanger and bent it back and forth really quickly? The metal heats up, possibly enough to burn you. The same thing happens when these moons flex. The change in pressure heats their interiors. It heats Io enough to actually melt its interior. Like the Earth, Io's molten insides break out of the surface in huge volcanoes. The first was discovered when the Voyager I probe cruised past the blighted moon in 1979. Many more have been found since then, and it looks like there are always volcanoes erupting on the poor moon.

The tidal friction also warms the other moons. Europa shows evidence of a liquid-water ocean buried under its frozen surface. That water may be heated by tides from passing moons.

If we look even farther out, we see more tides. Sometimes stars orbit each other in binary pairs. If the stars are very close together, tides can stretch them into egg shapes. If they are even closer, the stars can exchange material, passing streams of gas from one to the other. This changes the stars' evolution, affecting the way they age. Sometimes, if one of the stars is a dense, compact star called a white dwarf, the gas from the more normal star can pile up on the surface of the dwarf. When enough gas accumulates, it can suddenly explode in a cosmic version of a nuclear bomb. The explosion can rip the star to pieces, creating a titanic *supernova*, which can release as much energy in one second as will the Sun in its entire lifetime.

And we can take one more step out, to a truly grand scale. Whole galaxies are affected by tides, too. Galaxies, collections of billions of stars held together by their own gravity, sometimes pass close to each other. The differential gravity of one passing galaxy can not only stretch and distort but actually tear apart another galaxy. Sometimes, as with the binary stars, the more massive galaxy actually takes material—stars, gas, and dust—from the less massive one in an event called **galactic cannibalism**. This is hardly a rare event. There's evidence our own Galaxy has done this before, and as a matter of fact, we are currently colliding with a tiny galaxy called the Sagittarius Dwarf. It is passing through the Milky Way near the center, and as it does it loses stars to our much larger and more massive galaxy.

So the next time you're at the beach, think for a moment about what you're seeing. The force of tides may take the water in and out from the shoreline, but it also lengthens our day, pushes the Moon farther away, creates volcanoes, eats stars, and viciously tears apart whole galaxies. Of course, the tides also make it easier to find pretty shells on the coastline. Sometimes it's awesome to think about the universe as a whole, but other times it's okay just to wiggle your toes in the wet sand.

8

≈ ≈ ≈ ≈ ≈ ≈

The Moon Hits Your Eye
Like a Big Pizza Pie:
The Big Moon Illusion

On a warm spring evening when my daughter was still an infant, my wife and I put her in a stroller and set off for a walk through our neighborhood. Heading south, we turned onto a street that put us facing almost due west. The Sun was setting directly in front of us and looked swollen and flaming red as it sank to the horizon. It was spellbinding.

Remembering that the Moon was full that night, I turned around and faced east. There on the opposite horizon the Moon was rising, looking just as fat—though not as red—as the Sun, still setting 180 degrees behind us.

I gawked at the Moon. It looked positively *huge,* looming over the houses and trees, the parked cars and telephone poles. I could almost imagine falling into it, or reaching out and touching it.

I knew better, of course. I also knew something more. Later that evening, around 11:00 or so, I went outside. It was still clear, and I quickly found the Moon in the sky. After so many hours, the rotation of the Earth had carried it far from the horizon, and now the full Moon was bright and white, shining on me from high in the sky. Smiling wryly, I noted that the Moon appeared to have shrunk. From the vast disk glowering at me on the horizon earlier

that evening, the Moon had visibly deflated to the almost tiny circle I saw hanging well over my head.

I was yet another victim of what's called the Moon Illusion.

There is no doubt that the vast majority of people who see the
Moon rising (or setting) near the horizon think it looks far larger
than it does when overhead. Tests indicate that the Moon appears
about two to three times larger when on the horizon versus overhead.

This effect has been known for thousands of years. Aristotle
wrote of it in about 350 B.C., and a description was found on a
clay tablet from the royal library of Nineveh that was written
more than 300 years earlier than that date.

In modern popular culture there are many explanations offered
for this effect. Here are three very common ones: The Moon is
physically nearer to the viewer on the horizon, making it look bigger; the Earth's atmosphere acts like a lens, magnifying the disk of
the Moon, making it appear larger; and when we view the horizon
Moon we mentally compare it to objects like trees and houses on
the horizon, making it look bigger.

Need I say it? These explanations are wrong.

The first one—the Moon is nearer when on the horizon—is
spectacularly wrong. For the Moon to look twice as big, its distance
would have to be half as far. However, we know that the Moon's
orbit isn't nearly this elliptical. In fact, the difference between the
perigee (closest approach to the Earth) and apogee (farthest point
from the Earth) of the Moon's orbit is about 40,000 kilometers. The
Moon is an average of 400,000 kilometers away, so this is only a
10 percent effect, nowhere near the factor of two needed for the illusion. Also, the Moon takes two weeks to go from perigee to apogee,
so you wouldn't see this effect over the course of a single evening.

Ironically, the Moon is actually a bit *closer* to you when it's
overhead than when it is on the horizon, so it really appears bigger. The distance from the Moon to the center of the Earth stays
pretty much constant over a single night. When you look at the
Moon when it's on the horizon, you are roughly parallel to the line
between the Moon and the center of the Earth and roughly the
same distance away. But when you look at the Moon when it's

overhead, you are *between* the center of the Earth and the Moon. You're actually more than 6,000 kilometers closer to the Moon. This difference would make the Moon appear to be about 1.5 percent bigger when it's overhead than when it's on the horizon, not smaller. Clearly, the Moon's physical distance is not the issue here.

The second incorrect explanation—Earth's air distorts the Moon's image, making it look bigger—is also wrong. A ray of light will bend when it enters a new medium, say, as it travels from air to water. This effect is what makes a spoon look bent when it sticks out of a glass of water.

Light will also bend when it goes from the vacuum of space to the relatively dense medium of our atmosphere. As you look to the sky, atmospheric thickness changes very rapidly with height near the horizon. This is because the atmosphere curves along with the Earth (see chapter 4, "Blue Skies Smiling at Me," for an explanation). This change causes light to bend by different amounts depending on the angle of the light source off the horizon. When the Moon sits on the horizon, the top part is about a half a degree higher than the lower part, which means that light from the bottom half gets bent more. The air bends the light up, making it look as if the bottom part of the Moon is being squashed up into the top half. That's why the Moon (and the Sun too, of course) looks flattened when it sits directly on the horizon.

The vertical dimension is squashed but not the horizontal one. That's because as you go around the horizon, side to side, the thickness of the air is constant. It's only when the light comes from different heights that you see this effect.

Like the distance explanation, we see that near the horizon the Moon's disk is actually physically a little smaller than when it is high in the sky, so again this explanation must be wrong. Even so, this belief is commonly held by a diverse and widespread group of people. It's taught in high school and even in college, and I have heard that it is even used in textbooks, although I have never seen it in print.

Despite what your eyes and brain are telling you, if you go out and measure the size of the Moon when it is near the horizon and again when it is near the zenith, you will see that it is almost exactly the same size. You need not measure it accurately; you can simply hold a pencil eraser at arm's length to give yourself a comparison. If you do this, you'll see that even though the Moon *looks* huge near the horizon, you won't *measure* any difference.

The big Moon on the horizon effect is amazingly powerful. But the change in size is an illusion. So if this isn't a physical effect, it must be psychological.

The third explanation relies on psychology and doesn't need the Moon to be physically bigger; the Moon just has to be near other objects on the horizon. Mentally, we compare the Moon to these objects and it looks bigger. When it's near the zenith, we cannot make the same comparison, so it looks farther away.

But this can't be right. The illusion persists even when the horizon is clear, as when the Moon is viewed from ships at sea or out airplane windows. Also, you can position yourself so that you can see the zenith Moon between tall buildings, and it still doesn't look any bigger.

For further proof, try this: The next time you see the huge, full Moon on the horizon, bend over and look at the Moon upside-down from between your legs (you may want to wait until no one else is around). Most people claim that when they do this the effect vanishes. If the illusion were due to comparison with foreground objects, it would still persist while you were contorted like this, because even upside-down you could still see the foreground objects. But the illusion vanishes, so this cannot be the correct explanation either. Note, too, that this is further proof that the effect is not due to a measurable size change in the Moon's diameter.

So what *does* causes the Moon Illusion? I'll cut to the chase: no one knows, exactly. Although it's known positively to be an illusion, and it occurs because of the way our brains interpret images, psychologists don't know *exactly* why it occurs. There have been very firm claims made in the professional literature, but in my opinion the cause of the Moon Illusion is still not completely understood.

This doesn't mean we don't understand it at least partially. There are several factors involved. Probably the two most important are how we judge the size of distant objects and how we perceive the shape of the sky itself.

When you look at a crowded street scene, the people standing near you appear to be larger than the ones farther away. If you measured how big they looked by holding a ruler up near your eye and gauging the apparent size of the people around you, someone standing 5 meters (16 feet) away might look to be 30 centimeters (12 inches) tall, but someone twice as far away would look only 15 centimeters (6 inches) tall. The physical sizes of the images of these people on your retina are different, but you *perceive* them to be the same size. You certainly don't actually think the farther person is half the height of the nearer person, so somewhere in your brain you are interpreting those images, and you then think of the people as being roughly the same physical size.

This effect is called **size constancy**. It has clear advantages; if you actually perceived the more-distant people as being smaller, you would have a messed-up sense of depth perception. A species like that wouldn't survive long against a predator that knows very well just how far away (and how big) you are. In that sense, size constancy is a survival factor, and it's not surprising that it's a very strong effect.

However, we can be fooled. In the diagram on page 82 you see two lines converging to a point at the top. There are two horizontal lines drawn across them—one near the top where the lines converge, and the other near the bottom where they are farther apart. Which horizontal line is longer? Most people report the top one to be longer. However, if you measure them (and feel free to do so) you will see they are the same length.

This is called the Ponzo Illusion, after the researcher who characterized it. What's happening is that your brain is interpreting the converging lines to be parallel, like railroad tracks. Where they converge is actually perceived as being in the distance, just like railroad tracks appear to converge near the horizon. So your brain perceives the top of the diagram to be farther away than the bottom.

Now remember size constancy. Your brain wants to think that the top line is farther away. But since the length of the line is the same, your brain interprets this as meaning the top line is longer than the bottom line. Size constancy works in coordination with the perspective effect to trick your brain into thinking the upper line is longer when in fact it isn't.

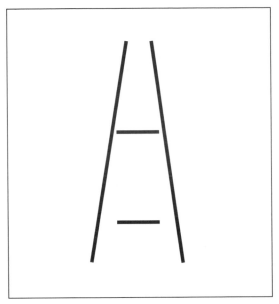

The Ponzo illusion is one of the most famous of all optical illusions. The horizontal lines are actually the same length, but the upper one appears longer because of the converging vertical lines.

What does this have to do with the Moon Illusion? For that we have to turn to the shape of the sky.

The sky is usually depicted in diagrams as a hemisphere, which is literally half a sphere. Of course, it isn't really; there is no surface above the Earth. The sky goes on forever. However, we do perceive the sky as a surface over us, and so it does appear to have a shape. In a sphere all points are equally distant from the center. The point on the sky directly overhead is called the zenith, and if the sky were indeed a sphere it would be just as far away as a point on the horizon.

But that's not really the case. Most people, myself included, actually see the sky as flattened near the top, more like a soup bowl than half a ball. Don't believe me? Try this: Go outside to level ground where you have a clear view of the sky from horizon to zenith. Imagine there is a line drawn from the zenith straight down across the sky to the horizon. Extend your arm, and point

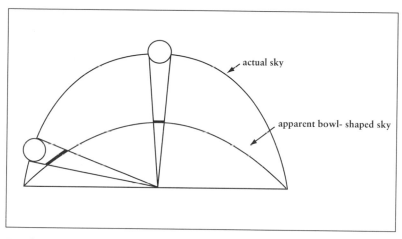

We don't see the sky as a hemisphere, but actually as a bowl inverted over our heads. When the Moon is on the horizon, it appears farther away than when it is overhead. Our brains are tricked into thinking the Moon is bigger than it really is when it's on the horizon.

your finger to where you think the halfway point is between the ground and the zenith, 45 degrees up from the horizon.

Now have a friend measure the angle of your arm relative to the ground. I will almost guarantee that your arm is at an angle of roughly 30 degrees and not 45 degrees, which is truly halfway up to the zenith. I have tried this myself with many friends (some of whom were astronomers), and no one has *ever* been higher than about 40 degrees. This happens because we see the sky as flattened; for a flat sky the midpoint between zenith and horizon is lower than for a hemispherical sky.

The reason for our perceiving the sky this way isn't well-known. An Arab researcher named Al-Hazan proposed in the eleventh century that this is due to our experience with flat terrain. When we look straight down, the ground is nearest to us, and as we raise our view the ground gets farther away. We interpret the sky the same way. This time as we look straight up, the sky appears closest to us, and as we then lower our gaze the sky appears farther away. Although this explanation is nearly 1,000 years old, it may indeed be the correct one.

But no matter what the cause, the perception persists. The sky looks flat. As Al-Hazan pointed out, this means that the sky looks farther away at the horizon than it does overhead.

Now we can put the pieces together. The Moon, of course, is physically the same size on the horizon as it is overhead. The shape of the sky makes the brain perceive the Moon as being farther away on the horizon than when it's overhead. Finally, the Ponzo Illusion shows us that when you have two objects that are the same physical size but at different distances, the brain interprets the more distant object as being bigger. Therefore, when the Moon is on the horizon, the brain interprets it as being bigger. The effect is very strong and has the same magnitude as the Ponzo illusion, so it seems safe to conclude that this is indeed the cause of the Moon Illusion.

This explanation was recently bolstered by a clever experiment performed by Long Island University psychologist Lloyd Kaufman and his physicist son, James, of IBM's Almaden Research Center. They used a device that allowed subjects to judge their perceived distance from the Moon. The apparatus projected two images of the Moon onto the sky. One image was fixed like the real Moon, and the other was adjustable in size. The subjects were asked to change the apparent size of the adjustable image until it looked like it was halfway between them and the fixed image of the Moon. Without exception, every person placed the halfway point of the horizon Moon much farther away than the halfway point of the elevated Moon, an average of four times farther away. This means they perceived the horizon as four times farther away than the zenith, supporting the modified Ponzo Illusion as the source of the Moon Illusion.

However, some people argue with this conclusion. For example, when you ask someone, "Which do you think is closer, the big horizon Moon or the smaller zenith Moon?" they will say the horizon Moon looks closer. That appears to directly contradict the Ponzo Illusion explanation, which says that the brain interprets the bigger object as *farther* away.

However, this isn't quite right. The Ponzo Illusion is that the farther-away object is bigger, not that the bigger object is farther away. See the difference? In the Ponzo Illusion the brain first unconsciously establishes distance and *then* interprets size. When you ask people which Moon looks bigger, they are first looking at size, and *then* consciously interpreting distance. These are two different processes, and may very well not be undertaken by the same part of the brain. This objection really has no merit.

In my opinion, the Ponzo Illusion coupled with size constancy and the shape of the sky is an adequate solution to the millennia-old Moon Illusion mystery. The real question may be why we perceive all these different steps the way we do. However, I am not a psychologist, just a curious astronomer. I'll note that as an astronomer, I am not fully qualified to judge competing psychological theories except on their predictions. It's quite possible that eventually a better theory may turn up, or that a fatal flaw in the Ponzo Illusion theory may arise. Hopefully, if that happens, the psychologists can explain it to astronomers so we can get our stories straight.

As an aside, I have often wondered if astronauts see this effect in space. One way or another, it might provide interesting clues about the root of the illusion. I asked astronaut Ron Parise if he has ever noticed it. Unfortunately, he told me, the Space Shuttle's windows are far too small to get an overview of the sky. Perhaps one day I'll see if NASA is willing to try this as an experiment when an astronaut undergoes a spacewalk. He or she could compare the size of the Moon when it's near the Earth's limb, its apparent outer edge, to how it appears when it is far from the Earth and see if the size appears to change. Interestingly, the experiment could happen much faster up there than here on Earth: the Shuttle's 90-minute orbit means they only have to wait 22 minutes or so between moonrise and when it's highest off the limb!

Having said all this, I'll ask you a final question: if you were to look at the full Moon and hold up a dime next to it, how far away would you have to hold the dime to get it equal in size to the full Moon?

The answer may surprise you: over 2 meters (7 feet) away! Unless you are extremely long-limbed, chances are you can't hold a dime this far away with your hand. Most people think the image of the Moon on the sky is big, but in reality it's pretty small. The Moon is about half a degree across, meaning that 180 of them would fit side by side from the horizon to the zenith (a distance of 90 degrees).

My point here is that often our perceptions conflict with reality. Usually reality knows what it is doing and it's we, ourselves, who are wrong. In a sense, that's not just the point of this chapter but indeed this whole book. Maybe we should always keep that thought in mind.

PART III

๛ ๛ ๛ ๛ ๛ ๛ ๛

Skies at Night Are
Big and Bright

If we dare journey beyond the Moon looking for bad astronomy, we'll find a universe filled with weird things waiting to be misinterpreted.

Meteors are a major source of bad astronomy. When two eighteenth-century Yale scientists proposed that meteors were coming from outer space, one wag responded, "I would more easily believe that two Yankee professors would lie than that stones would fall from heaven." That wag was Thomas Jefferson. Thankfully, he stuck to other things like founding the University of Virginia (my alma mater) and running the country, and steered clear of astronomy.

If you go outside on a cloudless night, you might see a meteor or two if you're lucky. If you are not too close to a city and its accompanying light pollution, you'll see hundreds or even thousands of stars. Like meteors, that starlight has come a long way; even the closest known star is a solid 40 trillion kilometers away. And like meteors, those stellar photons end up as so much fodder for our human misunderstanding of the cosmos. Stars have color, they twinkle, they come in different brightnesses, and all of these characteristics are subject to clumsy misidentification.

Bad astronomy can often force the doomsayers out into the open, too. This happened in the years, months, and days leading up to the "Great" Planetary Alignment of May 2000. Last I checked, the world had not ended. Cries of doom always seem to pop up at solar eclipses as well. Long heralded as omens of the gods' ill favor, eclipses are actually one of the most beautiful sights the sky provides.

Finally, in this section we'll travel back in time and space to where it all began, the Big Bang. Something about contemplating the beginning of everything twists our already tangled minds, and descriptions of the Big Bang usually confuse the issue more than unravel it. The irony of the Big Bang, I suppose, is that it is even odder than our oddest theories could possibly suppose.

9

Twinkle, Twinkle, Little Star: Why Stars Appear to Twinkle

"Twinkle, twinkle, little star, how I wonder what you are."
— Lyrics by Jane Taylor,
music by Wolfgang Amadeus Mozart

"Twinkle twinkle, little planet, can't observe so better can it."
— The Bad Astronomer

I was sitting at the observatory, waiting. It was 1990, and I was trying to make some observations as part of my master's degree work. The problem was the rain. It had poured that afternoon (not unusual for September in the mountains of Virginia), and I was waiting for the sky to clear up enough to actually get some good images.

After a few hours my luck changed, and the clouds broke up. Working quickly, I found a bright star and aimed the telescope there to focus on it. But try as I might, the image of the star on the computer screen never sharpened. I would move the focus in and out, trying everything, but no matter what I did the star image was hugely fuzzy.

So, I did what any astronomer locked in a small, dark room for three hours would do. I went outside and looked up.

The bright star I had chosen was high in the sky and twinkling madly. As I watched, it flashed spastically, sometimes even changing colors. I knew immediately why I couldn't get the star image sharp and crisp. The telescope wasn't to blame, our atmosphere was. I waited a couple of more hours, but the star refused to focus. I went home, resigned to try again the next night.

❧❧❧

Who hasn't sat underneath the velvet canopy of a nighttime sky and admired the stars? So far away, so brilliant, so . . . antsy?

Stars twinkle. It's very pretty. As you watch a star, it shimmers, it dances, it flickers. Sometimes it even changes color for a fraction of a second, going from white to green to red and back to white again.

But there, look at *that* star. Brighter than others, it shines with a steady, white glow. Why doesn't that one twinkle, too? If you wonder that aloud, a nearby person might smugly comment, "That's a planet. Planets don't twinkle, but stars do."

If you want to deflate them a little, ask them just *why* stars twinkle. Chances are, they won't know. And anyway, they're wrong. Planets can and do twinkle, as much as stars. It's just that twinkling rarely affects the way they look.

❧❧❧

Having an atmosphere here on Earth has definite advantages like letting us breathe, fly paper airplanes, spin pinwheels on our bikes, and so on. But as much as we all like air, sometimes astronomers wish it didn't exist. Air can be a drag.

If the atmosphere were steady, calm, and motionless, then things would be fine. But it isn't. The air is turbulent. It has different layers, with different temperatures. It blows this way and that. And that turbulence is the root of twinkling.

One annoying property of air is that it can bend a light ray. This is called **refraction,** and you've seen it countless times. Light bends when it goes from one medium to another, like from air to water or vice versa. When you put a spoon in a glass of water, the spoon looks bent where the air meets the water. But, really, it's just the light coming out of the water and into the air that bends. If you've ever gone fishing in a stream armed with just a net, you've experienced the practical side of this, too. If you don't compensate for refraction, you're more likely to get a netful of nothing than tonight's dinner.

Light will bend when it goes from one part of the atmosphere to a slightly less dense part. For example, hot air is less dense than cooler air. A layer of air just over the black tar of a highway is hotter than the air just above it, and light going through these layers gets bent. That's what causes the blacktop ahead of you to shimmer on a summer's day; the air is refracting the light, making the highway's surface look like a liquid. Sometimes you can even see cars reflected in the layer.

Here on the ground, the air can be fairly steady. But, high over our heads things are different. A few kilometers up, the air is constantly whipping around. Little packets of air, called **cells,** blow to and fro up there. Each cell is a few dozen centimeters across and is constantly in motion. Light passing in and out of the cells gets bent a little bit as they blow through the path of that light.

That's the cause of twinkling. Starlight shines steady and true across all those light years to the Earth. If we had no atmosphere, the starlight would head straight from the star into our eyes.

But we do have air. When the starlight goes through our atmosphere, it must pass in and out of those cells. Each cell bends the light slightly, usually in a random direction. Hundreds of cells blow through the path of the starlight every second, and each one makes the light from the star jump around. From the ground, the size of the star is very small, much smaller than the cell of air. The image of the star, therefore, appears to jump around a lot, so what we see on the ground is the star appearing to dance as the light bends randomly. The star twinkles!

Astronomers usually don't call this twinkling, they call it **seeing,** a confusing holdover from centuries past, but like most jargon, it's stuck in the language. Astronomers determine how bad the seeing is on a given night by measuring the apparent size of a star. A star's image dances around so quickly that our eyes see this as a blurring into a disk of light. The worse the seeing is, the bigger the star looks. On a typical night, the seeing is a couple of arcseconds. For comparison, the Moon is nearly 2,000 arcseconds across, and the naked eye can just resolve a disk that is about 100 arc-seconds across. The best seeing on the planet is usually a half an arcsecond, but it can be much larger, depending on how turbulent the air is.

Seeing also changes with time. Sometimes the air will suddenly grow calm for a few seconds, and the disk of a star will shrink dramatically. Since the light of the star gets concentrated into a smaller area, this lets you see fainter stars. I remember once sitting at the eyepiece of telescope for several minutes, looking for the very faint central star in a nebula. The star was just at the visibility limit of the telescope. Suddenly the seeing steadied up for a moment and the ghostly, pale-blue star snapped into my sight. Just as suddenly, the seeing went sour and the star disappeared. It was the faintest star I have ever seen with my own eyes, and it was amazing.

ৠ ৠ ৠ

So why don't planets twinkle? Planets are big. Well, in reality they're a lot smaller than stars, but they are also a *lot* closer. Even the biggest star at night appears as a tiny dot to the world's best tele-scopes, but Jupiter is seen as a disk with just a pair of binoculars.

Jupiter is affected by seeing just as much as a star. But, since the disk of the planet is big, it doesn't appear to jump around. The disk does move, but it moves much less relative to its apparent size, so it doesn't appear to dance around like a tiny star does. Small features on the planet are blurred out, but the overall planet just sits there, more or less impervious to turbulence.

More or less. Under especially bad conditions even planets can twinkle. After thunderstorms the air can be very shaky, and if the planet is on the far side of the Sun the planet's disk will look par-ticularly small, making it more susceptible to twinkling. But when a planet does twinkle, the seeing is incredibly bad, and observing is hopeless for that night.

Another way to increase twinkling is to observe near the hori-zon. When a star is just rising or setting, we are looking at it through more air because our atmosphere is curved. This means there are more cells between us and the star, and it can twinkle madly. Ironically, if you happen to be looking over a city, the air can be more stable. There are commonly smog layers over cities which stabilize the seeing, perhaps their only beneficial effect.

As it happens, different colors of light are refracted more easily than others. Blue and green, for example, bend much more than red. Sometimes, in really bad seeing, you can see stars change colors as first one color and then another is refracted toward you. Sirius is the brightest nighttime star, and it usually appears to be a steadily white color to the eye. But sometimes, when Sirius is low, it can flicker very dramatically and change colors rapidly. I have seen this myself many times; it's mesmerizing.

It can also lead to trouble. Imagine: You are driving along a lonely road at night and notice a bright object that appears to follow you. As you watch it flickers violently, going from bright to dim, and then you notice it's changing colors, from orange to green to red to blue! Could it be a spaceship? Are you about to be abducted by aliens?

No, you are a victim of bad astronomy. But the story sounds familiar, doesn't it? A lot of UFO stories sound like this. Stars appear to follow you as you drive because they are so far away. The twinkling of the star changes the brightness and the color, and imagination does the rest. I always smile when I hear a UFO tale like this one, and think that although it may not have been a UFO, it was *definitely* extraterrestrial.

❧❧❧

Twinkling stars may inspire songs and poetry, but astronomers consider them an inconvenience. One of the reasons we build big telescopes is that they help increase our **resolution** of objects. Imagine two objects, one of them half the size of the other, but both smaller than the seeing on a given evening. Because of seeing, they will both get blurred out to the same size, and we cannot tell which object is larger. This puts a lower limit on how small an object we can observe and still accurately measure its size. Anything smaller than this lower limit will be blurred out, making it look bigger.

Even worse, objects that are close together will get blurred together by seeing, and we cannot distinguish them. This really puts the brakes on how small an object we can detect.

There are actually several ways to work around seeing. One way is to work *over* it. If you launch a telescope up over the atmosphere it won't be affected by seeing at all. That's the basic reason the Hubble Space Telescope was put into orbit in 1990. Without an atmosphere between it and the objects it studies, it has a better view than telescopes on the ground (for more, see chapter 22, "Hubble Trouble"). Hubble is not limited by seeing, and can usually resolve objects much better than its land-based brethren. The problem is that launching a telescope is very expensive, and can make a space telescope cost ten times as much as one built on the ground.

Another way around seeing is to take a lot of really short exposures of an object. If the exposure is fast enough, it freezes the image of the star before the turbulent air can blur it. It's like taking a fast exposure of a moving object. A one-second exposure of a race car is hopelessly blurred, but one taken at $1/10{,}000$ of a second will be clean and clear. A very fast exposure will show a clear image of the star, but the position of the star's image will jump around from exposure to exposure as the light bends. Astronomers can take hundreds or thousands of very short exposures of a star and then add the separate images together electronically, yielding detail that is impossible with longer exposures. This technique was used to get the first resolved image of a star other than the Sun. The red giant star Antares was the target, and the image, though blurry, was definitely resolved and not just a point of light.

The big disadvantage of this technique is that it only works for bright objects. A faint one won't show up in the short exposure times necessary. This severely limits the available targets and therefore the usefulness of the process.

There is a third technique that shows amazing promise. If the observer can actually measure just how the atmosphere is distorting a star's image, then the shape of the telescope mirror itself can be warped to compensate for it. This technique is called **adaptive optics,** or AO for short, because the optical system of the telescope

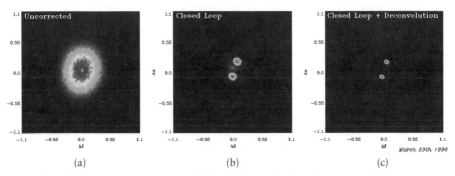

(a) (b) (c)

A close binary star pair may look like a blob of light when seen without adaptive optics (a), but is separated easily once the adaptive optics of the CFH telescope is switched on (b). Further image processing using computers can make the observation even better (c). The stars are separated by only about 0.3 arcseconds, or the apparent size of a quarter seen from a distance of almost 15 kilometers. (Image courtesy Canada France Hawaii Telescope Corporation, © 1996.)

can adapt to changes in the seeing. It's done by small pistons attached to rods, called **actuators,** located behind a mirror in the telescope. In some cases the rods push on the mirror, changing its shape, distorting the mirror just enough to correct for seeing changes. Another way is to use a collection of hexagonal mirrors that fit together like kitchen tiles, each with its own actuator. Little mirrors are much easier and less expensive to make than big ones, so many of the world's largest telescopes are designed this way.

The results are nothing less than incredible. The pictures above are from the Canada-France-Hawaii 3.6-meter telescope outfitted with AO. The image on the left is a picture of a binary star taken with the AO turned off. All we see is an elongated blur. But in the image on the right the AO is turned on, correcting for the seeing, and the two individual stars snap into focus.w

The European Southern Observatory has several telescopes in Chile outfitted with adaptive optics. One is the Very Large Telescope, or VLT for short. The name isn't exactly poetic, but it does describe the huge, 8-meter, hexagonally segmented mirror pretty accurately. There are actually four such 'scopes, and with adaptive optics their images rival Hubble's. One of the only disadvantages of adaptive optics is the narrow field of view; only a small area of the sky can be seen in each exposure. As the technology improves, though, so will the area, and eventually these telescopes will routinely use AO for much larger chunks of the sky.

❦ ❦ ❦

The next time you're out on a clear night and the stars dance their dance, you can remember how even the simplest things like the twinkle of a star can have complicated origins, and how difficult it can sometimes be to work around them.

Or, you could just watch the stars twinkle. That's okay, too.

10

Star Light, Star White: Stars of Many Colors

On a clear night, one of my favorite activities is to haul out my telescope and simply look at stuff in the sky. Usually, I have the 'scope set up in my yard, somewhere out of the way of trees, streetlights, and other obstacles. Still, a neighbor always manages to see me and drops by to take a peek.

The last time this happened, my neighbor brought her two school-age kids. They were being home-schooled and needed a science credit. She figured a night outside with a telescope would count.

After we looked at the Moon, Saturn, Jupiter, and a few other showpieces, the kids wanted to see a star through the telescope. I cautioned them that the stars would just look like points of light, and not disks. No ordinary telescope can magnify images that much. Then I turned the 'scope to Vega, one of the brightest stars in the sky. Without saying anything else, I let them take a look.

The gasps of delight were wonderful. "It's like a gem!" one of them breathed. "I can't believe how blue it is!"

I expected that reaction. My neighbor's daughter looked away from the 'scope and I pointed out Vega to her in the sky. She looked at it for a moment, and then said, "I didn't know stars really had color. I thought they were all white."

I expected that, too. I hear it a lot. Despite this common belief, stars *do* have color, and some are quite beautifully hued. Most look white because of our eyes; the fault lies with us and not the stars.

Amazingly, immense objects like stars emit colors because of the tiniest things of all: atoms.

Stars are basically giant balls of gas. Near the center, the immense pressure of the outer layers squeezes the atoms of gas together. When you squeeze something, it gets hot. The pressure is so high in the centers of stars like the Sun that the temperature can reach millions of degrees. At temperatures this high, the nuclei of atoms—their centers, composed of positively charged protons and neutral neutrons—smash into each other and stick together in a process called **nuclear fusion**. This process releases energy in the form of very energetic light called gamma rays.

Light acts like a messenger, transferring energy from one place to another; in this sense, light and energy are the same thing. The gamma rays don't get far before getting absorbed by another nucleus. They are promptly re-emitted, move out again, and get re-absorbed. This process happens over and over, countless trillions of times, and the energy of fusion in the center of the star works its way out to the surface.

When a gamma ray smacks into a subatomic particle, the particle increases its energy. In other words, it gets hot. Near the core the temperature can be millions of degrees, but the temperature drops with distance from the core. Eventually, near the surface of the star, the temperature is a comparatively chilly few thousand degrees Celsius (compare that to room temperature here on Earth, which is about 22 degrees Celsius).

This temperature is still more than enough to strip electrons from their parent atoms. All these particles near the Sun's surface are zipping around, bumping into each other, absorbing and emitting energy in the form of light. For a long time, it was a major problem in physics to figure out just how the Sun emitted this light. Around the year 1900, Max Planck, a German physicist, imagined that the particles in the Sun were like little oscillators, little vibrating springs. The mathematics of how oscillators give off energy is well understood, so he had grand hopes of figuring out how the Sun emits light.

But he couldn't get it to work. He assumed that the light was emitted in the form of a wave, and that each particle gave off a certain color of light. According to the physics of the time, any particle could emit any amount of energy it wanted. Planck, however, saw that this didn't really represent how a star emits light. He solved the problem by restricting the amount of energy each particle could produce. He realized that the emitted energy was *quantized,* meaning that the particles could only produce energy in even multiples of some unit. In other words, a star could give off 2 units of energy (whatever those units may be), or 3 or 4, but not 2.5, or 3.1. It had to be an integer, a whole number.

This was rather distasteful to Planck, who had no prior reason to assume this would be true. For centuries physicists assumed that energy flowed continuously, and not in tidy little bundles. Planck's model of quantized energy was flying in the face of all that. However, his model fit the data a lot better. He saw that it made the math work, so he published it.

This was how quantum mechanics was born.

Planck was right; light does come in a sort of minimum-energy packet. We call it a **photon.** Einstein used this idea in a paper about how light can eject electrons from metals, and he called it the **photoelectric effect.** Nowadays we use this effect to make solar panels, which provide power to devices from cheap calculators to the Hubble Space Telescope. Despite common belief, Einstein won his Nobel prize for this work and not his more celebrated work on relativity.

When Planck made his assumption about quantized energy, he found an interesting thing: the amount and color of the light a star emits depends on its temperature. If two stars were the same size, he determined, the hotter one would emit more light, and that light would be bluer than from the cooler star. Blue photons have more energy than red ones, and so a hotter star, with more energy, makes more energetic photons. A star at a certain temperature emits light at *all* different colors, but it emits most of its light at *one* specific color.

What this means is that a cool star, say around 2,500 degrees Celsius, emits its peak light in the red. A hotter star, near 6,000 degrees Celsius, peaks in the green. If the star gets even hotter, it emits mostly blue. Past that, the peak actually can occur in ultraviolet light, invisible to the naked eye.

That's the first key: the color of a star depends on its temperature. So, by measuring a star's color we can determine that temperature. The math for this is so well understood, as a matter of fact, that if we measure the amount of light a star gives off we can also determine how big it is. Amazingly, we can take a star's temperature and measure its girth just by looking at it! That's quite a feat given that the nearest star besides the Sun is 40 trillion kilometers (25 trillion miles) away.

However, just because a star's light peaks at a certain color doesn't mean it *looks* like it's that color. As an example, I give you the Sun: its color peak is actually in the green part of the spectrum, yet it looks white to us (go back and read chapter 4, "Blue Skies Smiling at Me," for more about our white Sun). The Sun gives off light at the blue and red ends of the spectrum as well, and it's the mix of all these colors that counts. Think of it this way: if I bake a batch of chocolate chip cookies, I put in more flour than anything else. Yet the cookies don't taste just like flour; they are a mix of all the other flavors, too. So it is with stars; the Sun emits more green light than any other one color, but it's the mix that makes the Sun white.

An interesting, and ironic, side note is that there are no intrinsically green stars. No matter what temperature a star is, the mixing of the colors guarantees that the overall color is not green. There are a couple of stars usually described as green by astronomers, but these are in binary systems; that is, they are very near another star. Usually, the other star is reddish or orange, and that can make something that is actually white look green in contrast. I have seen this myself; it's really weird to see a star glow green near its ruddy companion.

So if stars are all these different colors, why do most of them look white?

Look again. Which stars look white? If you start with the brightest stars in the sky, you may notice a clue: many of the brightest stars are blue-white or red. Sirius, the brightest star in the nighttime sky, is bluish. If you can see Sirius then perhaps Betelgeuse is also up, and it is quite orange. Antares, the heart of the constellation Scorpius, is rusty red. The name Antares means "rival of Mars" because their colors are so similar.

But as you go down the list, you'll notice that stars seem to lose their color as they get dimmer. Eventually, at some minimum brightness, all fainter stars look white. Clearly this is not something intrinsic to the stars, but to something inside of us.

That something is the construction of our eyes. We have two different kinds of cells in our retinae that detect light. **Rods** are cells that can determine the intensity, or brightness, of the light entering our eyes. **Cones** are cells that differentiate colors. (I used to get them mixed up, but now I think that *cones* see *colors*, which makes it easier). Rods are very sensitive and can even detect single photons if conditions are right. Cones, on the other hand, are just a tad dim of vision. They need to see lots of light before they can figure out what color it is. So, while a dim star may be bright enough for your rods to detect, allowing you to see the star, it may not be bright enough to trip your cones, and so you see no color. The star simply looks white. The star itself may be blue or orange or yellow, but there simply isn't enough light hitting your cones for them to figure out what's what.

That's a benefit of using telescopes of which many folks aren't aware. A telescope is more than an instrument used to magnify distant objects. A telescope collects light like a bucket collects rain. The bigger the bucket, the more rain you can collect. The bigger the telescope, the more *light* you collect. That light is redirected and focused into your eye, so even a faint star looks much brighter. Some stars that look white to the naked eye can be seen in their true color when viewed with a telescope. Even better, bright stars look even more colorful.

That's why the star Vega looked like a jewel to my astonished neighbor. Vega is the fourth-brightest star in the night sky, and it is one of the few to show color to the unaided eye. Take a peek through a telescope, though, and you see it in all its sapphire glory.

This brings me to a final thought. I love taking out my own telescope any old time, but the best night of all to do this is Halloween, when there are lots of kids around. Every year, my wife Marcella takes Zoe, our daughter, down the street trick-or-treating while I stay home. That way I can not only give out candy but also show the kids Jupiter or Saturn through the 'scope. Most of them have never looked through a real telescope before, and it's pretty nice to hear them exclaiming out loud when they see Saturn's rings.

I used to live in a fairly tough neighborhood, and some of the kids trick-or-treating looked like they were what teachers call high risk—prone to all sorts of problems, the least of which was dropping out of school. Yet these kids were the ones most likely to be shocked when they looked through my telescope and saw the moons of Jupiter. They would say, "Neat," or "Tough," or "Tight," or whatever the current jargon is for saying, "Wow!" Their cool exteriors were momentarily dropped when shown what the universe looks like up close.

A lot of people say that the current generation of children is bored and jaded. I heartily recommend that these people stop by an amateur astronomer's house some October 31. Maybe they'll see just how wrong they are.

Well, Well: The Difficulty of Daylight Star Sighting

I was never a boy scout.

That is probably a good thing. I was a smartass as a kid—some say that's still true—and I'm sure I would have had a hard time of it in the woods with just other boys my age for company. In high school I learned to be a bit of a practical joker, if only to exact revenge for some of the pranks pulled on me.

There is one trick traditionally done by Boy Scouts for which I'm sure I would have fallen. It's usually done late in the afternoon, while the Sun is still well up in the sky. It's best to do it after a long day in the woods, when everyone is exhausted and perhaps not thinking clearly. While sitting around resting, the discussion will turn to the Astronomy Merit Badge. One of the tests for the badge is constellation identification, so after a few minutes of talking, one of the boys (an older one) will get up and say, "Well, I need to practice finding constellations now."

This will, of course, raise some protest, usually by a tenderfoot. "But the Sun is still up," he'll inevitably say. "You can't see stars during the day!"

The older boy then puts on a condescending smile and says, "Of course I can. I just need to use my tube!" He then makes a tube out of rolled-up paper. Peering through it up at the sky, he'll make some comment like, "Ah, there's Orion now." He'll even invite other scouts (older boys, always) to take a look, and they all agree they can see some stars.

The young scout may resist for awhile, but, inevitably, curiosity will prevail. He'll ask to take a look. The older scout hands him the tube, which he obligingly puts up to his eye . . . and another scout then pours his canteen down in the tube, drenching the young victim.

That victim would certainly have been me. A skeptic and a loudmouth through and through, I would have vehemently protested any attempt to see stars during the day. I would also have been a wet kid.

The thing is, I would have been a *right* wet kid. Looking at stars through a tube during the day won't work. However, variations of this idea have been around a long time.

I've heard over and over again that it's possible to see stars in the daytime from the bottom of a tall chimney or a deep well. I've never heard a decent explanation as to *why* this should work, although people make vague claims about the brightness of the sky being greatly diminished in a deep well, making it easier to see stars. The sky is so bright it washes out the stars, they reason. By cutting back on the amount of skylight, stars are easier to see.

This idea certainly *sounds* reasonable. It also has a long history. The Greek philosopher Aristotle mentions it in passing in one of his essays. No less an author than Charles Dickens also endorsed it in at least one of his works. In his 1837 book, *The Pickwick Papers,* he opens his twentieth chapter with this tortuous sentence:

> In the ground-floor front of a dingy house, at the very farthest end of Freeman's Court, Cornhill, sat the four clerks of Messrs. Dodson & Fogg, two of his Majesty's attorneys of the courts of King's Bench and Common Pleas at Westminster, and solicitors of the High Court of Chancery—the aforesaid clerks catching as favourable glimpses of heaven's light and heaven's sun, in the course of their daily labours, as a man might hope to do, were he placed at the bottom of a reasonably deep well; and without the opportunity of perceiving the stars in the day-time, which the latter secluded situation affords.

Still awake? In other words, the clerks could see stars as easily as someone at the bottom of a well. Evidently, Dickens's publishers paid him by the word.

In a somewhat different version of this legend, Gregory of Tours, the sixth-century saint and historian, wrote in his *Libri Miraculorum* ("Book of Miracles") that the Virgin Mary drew water from a well, which became blessed by her presence. Those who are pious enough can gaze into the water from this well and, if they cloak their heads with cloths to block out the light from the sky, they see the Star of Bethlehem reflected in it. This is a rather neat trick: if you can't see it, you are not devout enough. Back to church with you!

The legend of seeing stars during the daytime is clearly tenacious, having been with us for a long chunk of our written history. I credit its longevity to the vague "scientificness" of the idea: as I pointed out before, it *sounds* like it might be true. Like eggs standing on end on the equinox, there is enough scientific jargon sprinkled in the legend that it bamboozles people. They don't understand it, so it must be true. The long history also lends support to it, but anecdotes are not conclusive proof! For that we need to turn away from hearsay and look to science.

Let's look closely at the legend: What is it about a chimney that might make it easier to see stars during the day? One obvious aspect is that it's dark at the bottom of a chimney. As your eyes get adapted to the dark, they become more sensitive to light. Perhaps that helps you to see stars.

Unfortunately, it won't work. Imagine you are sitting in the bottom of a tall chimney or smokestack, and it just so happens a star is directly overhead. Let's also imagine you have let your eyes get dark-adapted. But think about it for a moment: if your eyes have adapted to the darkness, and you are more sensitive to light from the star, the darkness also makes you more sensitive to the light from the *sky*. It won't be any easier to see a star. It's like standing in a loud bar talking to a friend. It's hard to hear him, so you use a hearing aid to increase your hearing sensitivity. But that won't work. You're focusing more sound in your ear from your friend, but you're also increasing the sound you hear from the rest of the bar. Nothing really changes, and it's just as hard to hear your friend.

Unfortunately, this also proves wrong the legend of seeing the Star of Bethlehem reflected from the water in a well. The water might reduce the brightness of the sky, but it reduces the brightness of the star by the exact same amount. You'd do better from the bottom of a chimney. That would change Nativity scenes extensively; a large smokestack next to the animals in the manger would take away a lot of the charm of Christmas.

You can see stars fairly easily at night, but not easily or at all during the day. The reason is just as obvious: at night, the sky is black and dark, but during the day it's very bright. The sky is bright during the day basically because the Sun lights it up. (See chapter 4, "Blue Skies Smiling at Me," for a more thorough explanation.)

The Sun isn't the only source of light illuminating the sky. If you go out at night during a full Moon, only the brightest stars will be visible, struggling to overcome the glaring light from the Moon. City lights also brighten the sky. This is called *light pollution,* and it's bad near cities, but it's not a good thing even near small towns. That's why astronomers try to build observatories far away from population centers.

During the day the bright sky swamps the rather meager light from the stars. As a matter of fact, on average the clear, daytime sky is roughly *six million* times brighter than that same patch of sky on a clear, moonless night. No wonder it's so hard to see stars during the daytime! They have to fight a fierce amount of light from the sky itself.

Still, we know it's possible to see the Moon, for example, during the day, so it's possible for some astronomical objects to be bright enough to be seen against the daytime sky. How bright must a star be before we can see it against the sky?

The critical item here is *contrast.* To see an object against a bright background, the object must be bright enough for your eye to pick it out over the rest of the light coming from all around the object. Tests done early in the twentieth century showed that the eye can pick out a star against the sky background if the object is roughly 50 percent as bright as the background. It may seem weird, at first, that you can see something that's fainter than the light around it. But the light from the star is concentrated in one spot, while the light from the sky is spread out all around it. The contrast with the sky is what makes the star visible.

Back in 1946, scientists performed experiments to see just how bright a star would need to be to poke out over the sky's glow. They mimicked what a human would see during the day as opposed to at night by tuning the amount of background light around an artificial star. They found that the dimmest star that a person could see during the day was about five times brighter than Sirius, the brightest star in the sky (besides the Sun). In other words, even the brightest star in the sky is too faint to be seen during the day (*Journal of the Optical Society of America* 36, no. 8 [1946]: 480).

Therefore, it's impossible for the unaided human eye to see *any* stars during the day. You'd think that's the end of the story, but there's still a twist to it. Those tests back in 1946 were done assuming the extra light was coming *from the entire sky*. If you are at the bottom of a chimney or a well, you aren't seeing the whole sky, just a little piece of it. If you can block out most of the glare from the sky, you *can* see fainter stars.

Very early in the twentieth century, two astronomers separately tried to figure out the eye's visibility limit, and to determine the faintest a star can be and still be seen against the night sky. They both found that by limiting the amount of sky they saw, they were able to greatly increase their ability to see faint stars. They determined that if you can cut out all but a tiny fraction of the sky, you can actually see stars that are about 10 times fainter than if seen in the whole sky—in which case it's *just* possible to see Sirius during the day, but that's it. The next brightest star, Canopus, is on the borderline of detectability. Let's be generous and say that both stars can be seen this way. Let's not forget, either, that there are bright planets visible to the naked eye: Mercury, Venus, Mars, and Jupiter can all appear brighter than Canopus or Sirius.

So we've determined that maybe, just maybe, we can just barely see six objects from a chimney, if the narrow opening of the chimney blocks the glare of most of the sky. We've done this by looking at all the advantages of viewing the sky from the bottom of a long, dark shaft. But we must be fair and look at the disadvantages as well.

There is one big one, and it's a deal-killer. Ironically, we looked at it as an advantage before: the narrow opening of the chimney. Before, it was good because it cut out glow from the sky, increasing contrast, making it easier to see stars. However, the small opening means there's less of a chance of a bright star passing into your field of view.

Most people think of the sky as being filled with stars. That's an illusion. You can see roughly 10,000 stars with the unaided eye, and they're spread out over the entire sky. We can estimate the average number of stars you might see through the opening at the top of a chimney. The answer may surprise you: even with a big opening, you will usually see only about 10 to 20 stars on the very darkest and clearest of nights. On a more typical night you might only see one or two stars. So, actually, looking through a chimney makes it a lot harder to see stars *even at night*. You are cutting out so much of the sky that only a few stars can be seen through the narrow aperture. During the day the odds are far, far worse. There are only six objects that you can see during the day to start with, not 10,000. The odds of one of these being in the chimney opening are remote indeed.

Scientists, of course, don't usually just calculate a number and assume it's correct. They actually go out and test it. An astronomer named J. Allen Hynek did just that and published his results in an issue of *Sky and Telescope* (no. 10 [1951]: 61). One day he took a few members of his astronomy class to an abandoned smokestack near Ohio University, where he taught. The bright star Vega—the fourth brightest in the sky—passes very close to directly overhead at that latitude, and they timed their experiment so that it would be in their field of view from the bottom of the smokestack. Vega is about half as bright as should be possible to see according to our calculations, but it is still one of the brightest stars in the sky. If it cannot be seen during the day, then certainly the vast majority of stars cannot be seen then, either.

At the appointed time Hynek and his students peered upwards, straining to see a glimmer from the star, but they all failed to observe it. Two students even used binoculars, which should have helped by increasing the contrast even more. They failed to see Vega as well. This is not surprising, really. Vega is too faint. Still, they showed by direct proof that stars are *at least* extraordinarily difficult to see through a chimney.

Another legend bites the dust, or in this case, the soot. While looking through a narrow opening does increase your ability to see faint objects, it simply doesn't increase it significantly enough to see stars during the day, and that same narrow opening makes it highly unlikely that a bright star will be in a viewable position.

Still, I have no doubts the legend will persist, as they all do. Even a friend of mine, an astronomer of no small status, swears the legend is true. He claims he saw it himself: he once looked up a long chimney during the day and saw a star. David Hughes, in his excellent paper entitled "Seeing Stars (Especially up Chimneys)," notes that a good chimney will have an updraft, even when there is no fire (*Quarterly Journal of the Royal Astronomical Society* 24 [1983]: 246–257). It's possible that my friend saw bits of debris caught in the draft and briefly lit by the Sun. At a great distance the debris will look tiny, unresolved, and not appear to move very quickly. This *could* be mistaken for a momentary glimpse of a star. I explained this to my friend, and I explained the idea of a star's brightness versus the sky's surface brightness, and I even talked about the odds of a bright star just happening to be in that extremely tiny line of sight, but he would have none of it. He stands by his story. I guess even the staunchest of scientific minds can have superstitions they don't want to drop. It's an interesting cautionary tale for all of us, I think.

Now, having said all that, I must confess that it *is* possible to easily see one starlike object during the day: Venus. Venus is roughly 15 times brighter than Sirius, so not only is it possible to see during the day, it's also relatively easy. You need to know just where to look, but it can be done. I've seen it myself on several occasions, in broad daylight. However, extrapolating from seeing Venus during the day to seeing stars from a chimney is a pretty big stretch. In the end, the legend turns out to be just that: a legend.

❧ ❧ ❧

A final note on this topic: I know for a fact that I would fall for that old boy scout tube trick. Why? Because a variation got me when I was about seven or eight years old, except I was told it was a coordination test. I was supposed to roll up a paper plate, put it in the front of my pants so that it stuck out a few centimeters, balance a rock on my nose, and then tilt my head forward so that the rock fell into the rolled-up paper plate.

As soon as I tilted my head back one of the other kids poured a glass of *ice cold* water into the paper-plate tube. This incident may have mentally scarred me for life; I still shrink away from picnics featuring paper plates. For all I know, the stunt gave me a core of vehemence against such things, which in turn led to the book you are holding in your hands right now. So, I say to those older kids who played such a mean trick on a naive young kid: thanks!

12

❧❧❧❧❧❧

The Brightest Star: Polaris—
Just Another Face in the Crowd

A few years ago I was chatting with a friend of mine. The night before, he claimed to have seen a bright, slowly moving object in the sky. I realized immediately that he had seen a man-made satellite, but his description confused me. The problem was the *way* he described where it was in the sky. He said the object was in the west, near the horizon, but he also said it was near Polaris.

"But Polaris isn't in the west," I told him. "It's in the north. And it's well above the horizon."

"Oh, well, the thing I saw was near this really bright star just after sunset," he replied.

Aha! I thought. The bright "star" must have been the planet Venus, which was low in the western sky at dusk at that time of year. Venus was almost painfully bright, far brighter than any other star in the sky, brighter even than most airplanes. He thought it was Polaris; and when I finally figured all this out I realized I had stumbled onto some more bad astronomy.

A lot of people think Polaris is the brightest star in the sky. Let's get this right off the bat: it isn't. Polaris just barely makes it onto the list of the top 50 brightest stars, and, as a matter of fact, it is hard to see if you live in even moderately light-polluted skies. Growing up in suburban Washington, D.C., I could barely see it. If the sky was even a little hazy, which it often is on the east coast of the United States, I couldn't see it at all.

Okay, so Polaris is a bit of a dim bulb. Why, then, is it often mistaken for a powerhouse? I have a theory: people confuse *brightness* with *importance*.

Polaris isn't a bright star, but it is an important one. The reason it's important is that it sits very close to the sky's north pole. And to see just why the sky has a north pole at all, we need to do something we've already done a few times in this book: start with the Earth beneath our feet.

The Earth is basically a giant ball. It's also a *spinning* ball. A sphere sitting all by itself has no real up or down. Nothing on its surface is any different than any other part. But when you spin it, it automatically gets two points that are easily defined: the points where the spin axis intersects the surface. On the Earth, we call these the **north** and **south poles.** By definition, the north pole is the point at which, if you are above it looking down, the planet appears to spin counterclockwise. Another interesting place is the line that goes around the Earth halfway between the poles; this is the **equator.**

Of course you've heard this before, but now comes the fun part. We observe the sky from the Earth, and even though the sky itself isn't spinning, to us it looks like it does because *we* are spinning. We think of the Sun and the stars as rising and setting during the day and night, but really we are the ones turning around on our giant spinning ball, not the sky. Still, it's convenient to think of the sky as spinning. Ancient astronomers thought the stars were holes in a giant sphere through which shone the light of heaven. Nowadays we know better, but it's still a useful model.

Imagine the sky really is a ball spinning around us. Just like the Earth, then, it has a north pole and a south pole, which we call the **north celestial pole,** or NCP for short, and the **south celestial pole** (SCP), to distinguish them from the ones on the Earth. They are reflections of the Earth's own features on the sky. If you were to stand on the Earth's north pole, the north celestial pole would appear to be straight up, directly over your head. The south celestial pole would be straight down, beneath your feet, where you can't see it—there's 13,000 kilometers of spinning planet in the way.

Let's stay at the north pole for awhile (I hope you're dressed warmly). It's nighttime, and you watch the stars. As the Earth turns under your feet, you'll see the sky turn above you. All the stars will appear to make circles over the course of a 24-hour day. Stars near the NCP will make little circles, and stars near the horizon make big ones. All these circles will be centered on the point straight over your head: the NCP.

Can't picture it? Then stand up! Really. Find a room with an overhead lamp, or something in the ceiling you can stand under and use as a reference point. Once there, start spinning, slowly— if you get dizzy you won't be able to read the rest of this chapter. See how the point over your head stays put while you spin? That's because it's your own private NCP. Look at the windows: they appear to make big circles around you as you spin, but that dead spider near the light that you've been meaning to vacuum out for a month appears to make only a little circle.

So it is with the sky. Stars near the NCP make little circles, and stars far from it makes big ones. The NCP takes on a special importance, because all the stars in the sky look like they circle around it. This is true for anywhere on the Earth from which the NCP is visible; that is, anywhere north of the equator. These same arguments are true as well for the SCP. An important thing to know is that since the Earth is spinning, and not just yourself, no matter where you are, the stars go around the NCP while the NCP always hangs in the same spot in the sky. It's like the Earth's axis is a giant arrow, and at the north pole it sticks out of the Earth and always points to the same position in the sky. It's always in the north because no matter where you are on the Earth, the north pole is to the north.

Remember, these places on the sky are just like places on the Earth, but projected into the sky. For me, it's behind an ancient maple tree when I look at the sky from my backyard. For you, it might be next to a building, or over a mountain, or beneath the ledge of the apartment above yours; but *it's always there*. It never moves.

A long exposure of the night sky reveals the elegant motion of the stars. From our vantage point on the spinning Earth, the stars appear to make circles in the sky. In this picture, taken in Colorado and facing north, the stars in the northern hemisphere arc around Polaris. Note that Polaris is not exactly on the pole, so it too makes a short arc. (Photograph courtesy of Jon Kolb, Adventures in Astrophotography, http://home.datawest.net/jkolb/.)

Now, as it happens, there is a middling bright star near the NCP. You wouldn't give it a second glance if it were anywhere else on the sky, but since this one is near the NCP it never rises and it never sets. All night long this star sits there while other stars get higher or lower in the sky. Wouldn't you think it's important? Think of it this way: before people had satellites, or airplane reconnaissance, or handheld Global Positioning System (GPS) devices, they had to know north from south and east from west. This star took on great importance to them because it showed them which way was north, all night long. Even today, if you get lost in the woods without a compass you'll be glad to see it.

This star has the somewhat unremarkable name of Alpha Ursa Minoris, but due to its proximity to the NCP it has taken on the popular name of Polaris. The star itself is actually rather interesting; it's really a multiple star consisting of at least six stars in orbit around each other. They appear to be one star to us because they are so far away—430 light-years—that all the stars merge into one point of light, the same way that a pair of headlights on an automobile might look like one light from far away.

Polaris is hundreds of light-years away, so the fact that it's near our NCP is simply a coincidence. Just to prove that point, the nearest star to the *south* celestial pole is the barely visible star Sigma Octans, which is something like the three-thousandth brightest star in the sky. And note that these stars only work for the Earth; from another planet, like Jupiter, Polaris is nowhere near its NCP.

Actually, it's not even precisely on the NCP as seen here on Earth. Currently, Polaris sits about a degree away from the NCP, the equivalent to twice the diameter of the full Moon as seen from the Earth. Still, compared to the whole sky, that's pretty close.

But it's more than just a coincidence in space; it's a coincidence in *time* as well.

Remember, Polaris is what it is because the Earth's axis points more or less toward it. However, the Earth's axis isn't perfectly fixed in space. As we saw in chapter 5, "A Dash of Seasons," the Earth's axis drifts slowly in space, making a circle roughly a quarter of the sky across every 26,000 years or so. This *precession of the axis* means that the Earth's north pole changes its position relative to the sky over time. So the fact that it's near Polaris right now is simply a coincidence. Over the years the Earth's pole will move slowly away from Polaris, leaving behind the relatively faint star, demoting it to its proper place among the second-tier stars in the sky.

Worse, in 14,000 years or so, the star Vega will be near the NCP. Vega is the fourth-brightest star in the sky, a shining, brilliant-blue gem in the northern summer sky, and very obvious even in light-polluted skies. If people mistake the brightness of a star with its importance now, with the dim Polaris sitting on the throne, then the situation will be far worse when Vega occupies that position.

Until that time off in the distant future, we'll still need Polaris to tell us which way is north, and that's enough to make Polaris important. But it's still not bright, which is why I think people confuse its brilliance—or lack thereof—with its stellar status. Just like people, stars can be important without being terribly bright.

13

Shadows in the Sky:
Eclipses and Sun-Watching

We humans have spent a long, hard time learning that the Earth is not a special place. It's not at the center of the universe, there are probably millions of planets like it in the Galaxy, and we may not even be the only place with life.

But there is one thing special about our blue home. It's a coincidence of place as well as time, and it is unique among all the moons and planets in the solar system. The Sun is much bigger than the Moon—about 400 times as big—but it's also 400 times farther away from us. These two effects cancel each other out, so, from our perspective down here on Earth, the Moon and the Sun appear to be the same size in the sky.

Normally, you'd hardly notice this. For one thing, the Sun is so bright it's hard to look at, making its size difficult to judge. For another, when the Moon and Sun are near each other in the sky, the Moon is a thin crescent and difficult to see. (Check out chapter 6, "Phase the Nation," for more about the Moon's position and shape relative to the Sun.)

But there is one time when it's pretty obvious that they're the same size, and that's when the Moon passes directly in front of the Sun. When that happens, the Moon blocks the Sun and we get what's called a **solar eclipse**. The eclipse starts small, when only a bit of the Sun gets blocked by the edge of the Moon. But as the

Moon's orbital motion sweeps it around the Earth, more and more of the Sun disappears behind the Moon's limb. We see the Moon in silhouette, a dark circle slowly covering the Sun. Eventually, the entire disk of the Sun is blocked. When this happens, the sky grows deep blue, almost purple, like at sunset. The temperature drops, birds stop singing, crickets will chirp, and it's like having a little night in the middle of the day.

This would be odd enough, but at the moment of totality, when the Sun's disk is completely covered by the Moon, the Sun's outer atmosphere, called the **corona**, leaps into view. Normally invisible because the Sun's surface is vastly brighter, the corona is wispy, ethereal, and surrounds the Sun like a halo or aura. When the corona becomes visible, viewers almost universally gasp in awe and delight, and some have been brought to tears by the sheer beauty of it.

Eclipses are magnificent, and they do not happen very often, but they are predictable. The Moon's path in the sky has been charted for millennia, and ancient astronomers could predict eclipses with perhaps surprising accuracy. It's not surprising then that historical records are full of tales about eclipses. Mark Twain even used one in his novel *A Connecticut Yankee in King Arthur's Court*. In it a young man from America is transported back in time to medieval England and, through a variety of circumstances, ends up being sentenced to be burned at the stake. However, he happens to know that a total solar eclipse is about to occur and tells his captors that if they don't release him, he'll take away the Sun. Of course, the eclipse happens right on schedule and he is set free.

That may sound silly, but it's based on an actual event, and none other than Christopher Columbus is in the leading role. In 1503, on his fourth voyage to America, Columbus was stranded in Jamaica, his ships too damaged to be seaworthy. He relied on the natives for food and shelter, but they soon became weary of feeding Columbus's men. When the natives told him this, Columbus remembered that a lunar eclipse—when the Earth's shadow falls on the Moon, turning it dark—would occur soon. Just as Twain retold the tale nearly 400 years later (with a solar instead of a lunar eclipse), the event terrified the natives, who then begged Columbus to bring back the Moon. He did, and he and his men were able to stay on the island until they were rescued.

Occasionally, events are antithetical to the Columbus story— *not* predicting an eclipse can get you in trouble. In ancient China it was the duty of all court astronomers to predict solar eclipses. The Chinese thought that a solar eclipse was a giant dragon eating the Sun, and if enough advance warning were given they could chase the dragon away by beating drums and shooting arrows in the sky. In 2134 B.C., as the story goes, two hapless (and perhaps apocryphal) astronomers by the names of Hsi and Ho didn't take their duties too seriously. They knew a solar eclipse was coming, but decided to hit the tavern first before telling the emperor. They drank too much and forgot to pass on the news. When the eclipse came, everyone was caught off guard. Luckily, the emperor was able to "scare off" the dragon, and the kingdom was saved. Hsi and Ho weren't so lucky. They were collected, thrust before the emperor, presumably chastised, and not-so-presumably had their heads cut off. Legend has it that the emperor threw their heads so high in the air that they became stars, which can be dimly seen between the constellations of Perseus and Cassiopaeia. (Today we know these two faint objects to be clusters of thousands of stars, and they're a pretty sight through a small telescope—prettier, no doubt, than this gruesome Chinese legend might have you think.) The lesson here is still relevant today: publish first, and *then* head off—so to speak—to the bars.

Even today, people are superstitiously terrified of eclipses. After a total solar eclipse in August 1999 that was seen all over Europe, I had an e-mail conversation with a young woman from Bosnia, which was then suffering from terrible fighting. She was shocked and saddened to see the streets deserted during the eclipse and the signs posted to warn of dangerous rays from the Sun that would kill people exposed to them. As if these people didn't have enough to worry about, they also had to hide in fear of something that might have actually given them a fair degree of much-needed joy.

Not all eclipse fears are so severe. There are many legends among ancient people about solar eclipses, and having the Sun eaten is a common thread. Others have seen it as a bad omen, so they pray during eclipses. Still others avert their eyes, lest they have a spell of bad luck cast over them . . .

. . . which brings us to a very interesting and somewhat controversial point about the Sun and eclipses. How many times have you heard that looking at an eclipse will make you go blind? Every time a solar eclipse rolls around, the news is full of warnings and admonitions. The problem is, they never say exactly *why* you can go blind, or what degree of eye damage you might suffer. Worse, they sometimes give incorrect advice on viewing an eclipse, increasing the danger.

I'll cut to the chase: viewing an eclipse can indeed be dangerous. Obviously, looking at the Sun is very painful, and it is extremely difficult to do so without flinching, tearing up, or looking away. The Sun is just too bright to look at. Every astronomy textbook I have ever read has an admonition against looking directly at the Sun, and it is common knowledge that looking at the Sun, even briefly, can cause permanent damage to your eyes.

While researching information on solar eye damage for this chapter, I stumbled across an amazing irony: while it *is* dangerous to view an eclipse with the unaided eye, it is actually far *less* dangerous to look at the Sun when it is *not* eclipsed! This may sound contradictory, but it actually is due to the mechanisms inside the eye that prevent overexposure from light.

There is copious evidence that little or no long-term damage results from observing the uneclipsed Sun. I was shocked to find this information; I have been steeped in a culture that says looking at the Sun with the unaided eye will result in permanent and total blindness. However, this is almost certainly not the case.

Andrew Young, an adjunct faculty member of the San Diego State University's Department of Astronomy, has collected an astonishing amount of *misinformation* concerning solar blindness (see http://mintaka.sdsu.edu/GF/vision/Galileo.html). His research flies in the face of almost all common knowledge about solar blindness. He is quite strong in his statement: under normal circumstances, glancing at the Sun will not permanently damage your eyes.

"The eyes are just barely good enough at rejecting [damaging] light," Young told me, because the pupil in the eye constricts dramatically when exposed to bright light, cutting off the vast majority of light entering the eye. Most people's retinae don't get overexposed when they glance at the Sun. Young quotes from a paper, "Chorioretinal Temperature Increases from Solar Observation," published in the *Bulletin of Mathematical Biophysics* (vol. 33 [1971]: 1–17), in which the authors claim that under normal circumstances, the constriction of the eye's pupil prevents too much light from the Sun from actually damaging the retina. There may be a slight (4 degrees Celsius) temperature rise in the tissue, but this is most likely not enough to cause permanent damage.

However, natural variations in the amount of pupil constriction between different people means that some might still be prone to retinal damage this way. These people make up the majority of **solar retinopathy** patients—people who suffer eye damage from looking at the Sun.

According to physicians at the Moorsfields Eye Hospital in London, England, observing the Sun can cause damage to the eye but not total blindness. On their web site [http://www.moorfields.org.uk/ef-solret.html], they report that half their patients with eye injury recover completely, only 10 percent suffer permanent vision loss, and, most interestingly, never has anyone had a total loss of vision from solar retinopathy.

So there is damage, and sometimes it can be severe, but most people recover, and no one has ever become totally blind by looking at the Sun. However, due to the natural variation in pupil constriction from person to person, I think I still need to stress that while it very well may be safe (or at least not very dangerous) to glance at the Sun, staring at it may still cause damage. The damage is most likely minimal, but why take chances? Try not to stare at the Sun, and try to minimize any glances at it. You might be part of the group that will suffer some injury from it.

So, if the full Sun is not likely to be dangerous, why should viewing a solar eclipse cause eye injury? During an eclipse most or all of the Sun is blocked by the Moon. However, think about what happens inside your eye when you view an eclipse. During a total eclipse the Sun's surface is completely covered by the Moon, and the sky grows dark. When this happens, *your pupil dilates*; that is, *it opens up wide*. It does this to let in more light so you can see better in the dark.

The Moon completely blocks the Sun's disk for a few minutes at most. Suddenly, when this phase of the eclipse ends, a small sliver of the Sun is revealed. Even though the *total* light from the Sun is less than when it is not being eclipsed, *each part* of the Sun is still producing just as much light. In other words, even if you block 99 percent of the Sun's surface, that remaining 1 percent is still pretty bright—it's 4,000 times brighter than the full Moon. An eclipse is not like a filter, blocking the light from hitting your eyes. Any piece of the Sun exposed will still focus this harmful light onto your retina, causing damage.

So when the Sun becomes visible again, with your pupil dilated wide, all that light gets in and hits your retina—and it's then that sunlight can really and truly hurt your eye. The bluer light can cause a photochemical change in your retina, damaging it, although most likely not permanently. This effect, according to Young, is worse in children because as we age the lenses in our eyes turn yellowish. This blocks blue light, better protecting older retinae from the damage. However, children's lenses are still clear, letting through the bad light. So while it's dangerous to look at an eclipse, it's even more so for children.

I'll note here that another common misconception about an eclipse is that the x-rays emitted from the Sun's corona can damage your eyes. The corona is extremely hot, but so tenuous that the normally bright Sun completely overwhelms it, making it extraordinarily difficult to observe during the daytime.

The corona is so hot that it gives off x-rays. Most folks know that x-rays are dangerous; after all, you have to wear a lead shield when getting an x-ray at the dentist. So many people put these two facts together and assume that it's the corona that can damage your eyes during an eclipse.

This is wrong. X-rays from any source in the sky cannot penetrate the Earth's atmosphere, which, for all intents and purposes, absorbs every x-ray photon coming from space. It acts like a shield, protecting us. Even if the atmosphere were not here, the corona is also simply too faint to hurt our eyes. And remember: whether the Sun is eclipsed or not, the corona is still there; it's just too faint to see. So if it could hurt your eyes during an eclipse, it could do so at any random time. In reality, the corona can't hurt you.

✎✎✎

There are several ways to enjoy an eclipse without risking your eyes. You can use a telescope or binoculars to project the image of the Sun onto a piece of paper or a wall. You can wear very dark goggles, like welders wear; make sure they are rated as #14 so that they are dark enough to be comfortable.

You can also use a solar filter on a telescope or binoculars, but only the kind that mounts in front of the main lens or mirror. This stops most of the light from entering the optics in the first place. Some companies sell filters that go on the eyepieces, which block the amount of light leaving the optics. However, the optics focus all that sunlight right onto that filter, which can heat up a lot. Filters like this have been known to melt or crack. I heard one story of a solar filter that actually exploded! That's bad enough, but then your eyes are flooded with all that sunlight concentrated by the optics. The lesson: stay clear of such devices.

Also, despite some advice I have seen, do not use unexposed film to block the light. Even as lofty a source as the CNN web site once claimed that it was safe to view an eclipse this way. This is actually a very dangerous way to do it; it lets through less visible light, so your pupils widen. However, it does not block the dangerous wavelengths of light, so that even more damaging light floods your eye. I and several hundred other people flooded CNN with e-mail, and the web site was hastily fixed.

Solar eclipses do not happen very often, and they usually occur over scattered parts of the planet. I have never seen a total solar eclipse, though I've seen a dozen or so partial ones. Someday I hope to see a total eclipse myself, but when I do, I'll be careful.

And I'd better hurry. As is discussed in chapter 7, "The Gravity of the Situation," the Moon is slowly receding from the Earth. It's only moving away about 4 centimeters (2 inches) a year, but over time that adds up. As it gets farther away, it appears smaller in the sky. That means eventually it will be too small to completely cover the Sun during a solar eclipse. Instead, we'll get an **annular** eclipse, a solar eclipse where the Moon's size is somewhat smaller than the Sun, and you can see a ring of Sun around the dark disk of the Moon. We get these eclipses now because the Moon's orbit is elliptical, and if the eclipse occurs when the Moon is at the highest point in its orbit, the eclipse is annular. But, eventually, these will happen all the time. The corona will forever be hidden by the glare of the Sun and solar eclipses will be interesting, but lack the impact they have now. That's why, in the beginning of this chapter, I said that total solar eclipses are a coincidence of space and time. Given enough time, they won't happen anymore.

ର ର ର

I can't leave this chapter without busting up one more misconception. It is an extremely common story that Galileo went blind because he observed the Sun through his telescope. I have said this myself, even once on my web site. Andy Young e-mailed me about it and set me straight.

Galileo did indeed go blind. However, it was *not* due to observing the Sun. Galileo realized rather quickly that looking through his small telescope at the Sun was a quite painful experience. Early on he only observed the Sun just before sunset, when it is much dimmer and safer to see. However, he later used a projection method to view the Sun and observe sunspots. He simply aimed his telescope at the Sun and projected the image onto a piece of paper or a wall, casting a much larger image of the Sun. This method is far easier and produces a large image that is easier to study as well.

Certainly, using a telescope to observe the Sun can indeed cause damage to the eye, since a telescope gathers the sunlight and concentrates it in your eye (much the same way that you can burn a leaf with a magnifying glass). However, this sort of damage occurs very rapidly after solar observation, and Galileo did not go blind until he was in his 70s, decades *after* his solar observations. There is copious documentation that during the intervening years his eyesight was quite good. Galileo suffered from cataracts and glaucoma later in life, but this was clearly not from his telescopic observations.

Galileo's observations of sunspots on the Sun caused quite a stir; the Catholic church had considered the Sun to be unblemished and perfect. Together with his observations of Jupiter, Venus, the Moon, and the Milky Way itself, he revolutionized our way of seeing and thinking about science, ourselves, and our universe. Yet we still can't get simple stories about him correct. Maybe *we* are the ones who are sometimes blind.

14

The Disaster That Wasn't: The Great Planetary Alignment of 2000

On May 5, 2000, the Earth was not destroyed.

Perhaps you missed this, since you were busy living, eating, going to work, brushing your teeth, etc. However, in the months before May 5, 2000, a lot of people actually thought the Earth would be destroyed. Instead of the usual culprits of nuclear war, environmental disaster, or the Y2K bug, this particular brand of global destruction was to have been wrought by the universe itself, or, at least, our small part of it.

On that date, at 8:08 Greenwich Mean Time, an "alignment of the planets" was supposed to have caused the Final Reckoning. This Grand Alignment—also called the "Grand Conjunction" by the prophesiers of doom to make it sound more mysterious and somehow more *millennial*—would throw all manner of forces out of balance, causing huge earthquakes, a possible shift in the Earth's poles, death, destruction, higher taxes, and so forth. Some even thought it would cause the total annihilation of the Earth itself. The tool of this disaster was to have been the combined gravity of the planets in the solar system.

These people, obviously, were wrong. Some of them were honest and simply mistaken, others were quacks and didn't know any

125

better, and still others were frauds trying to make money off the misinformed. Nonetheless, they were all wrong, plain and simple.

Of course, there's a long and not-so-noble history of misinterpreting signs from the sky. Long before studying the skies was a true science, there was astrology. Astrology is the belief that—contrary to every single thing we know about physics, astronomy, and logic—somehow the stars and planets control our lives. The reason astrology came about is not so hard to understand. People's lives can seem to be out of their control. Capricious weather, luck, and happenstance seem to influence our lives more than we can ourselves. It's human nature to be curious about the causes of such things, but it's also human nature to pervert that curiosity into blame. We blame the gods, the stars, the shaman, the politicians, everyone but ourselves or simple bad luck. It's natural to try to deny our own involvement and wish for some supernatural causation.

There *is* some connection between what happens in the sky and what happens down here on Earth. Agriculture depends on weather, and weather depends on the Sun. Agriculture also depends on the seasons, and these can be predicted by watching the skies. In winter, the Sun is lower in the sky during the day and it is not up as long. Certain constellations are up when it's cold, and others when it gets hot. The sky and the Earth seem irrevocably connected. Finding patterns in the sky that seem to have spiritual puppet strings tied to us here on Earth was perhaps inevitable.

Eventually, everything in the sky, from comets to eclipses, was assumed to portend coming events. It may be easy to laugh off such superstitions as the folly of simple people from ancient times. However, even today, firmly into the twenty-first century, we still deal with ancient superstitions that we simply cannot seem to cast off. Just a few months into the new century we had to deal with yet another instance of the shadow of our primitive need to blame the skies. The May 2000 planetary-alignment-disaster-that-wasn't spawned a whole cottage industry of gloom and doom, but, like all signs from the sky throughout history, it turned out to be just another false alarm. As with most superstitions, the rational process of the scientific method came to the rescue. To find out how, let's take a look at what an "alignment" really is.

All the planets in the solar system, including the Earth, orbit the Sun. They move at different speeds, depending on how far they are from the Sun. Tiny Mercury, only 58 million kilometers from the Sun, screams around it in just 88 days. The Earth, almost three times as far, takes one full year—which is, after all, how we define the year. Jupiter takes 12 years, Saturn 29, and distant, frigid Pluto 250 years.

All the major planets in the solar system formed from a rotating disk of gas and dust centered on the Sun. Now, nearly 5 billion years later, we still see all those planets orbiting the Sun in the same plane. Since we are also in that plane, we see it edge-on. From our vantage point, it looks like all the planets travel through the sky nearly in a line, since a plane seen edge-on looks like a line.

Since all the planets move across the sky at different rates, they are constantly playing a kind of NASCAR racing game. Like the hands of a clock only meeting every hour, the swifter planets can appear to "catch up" to and eventually pass the slower-moving ones. The Earth is the third planet out from the Sun, so we move in our orbit faster than Mars, Jupiter, and the rest of the outer planets. You might think, then, that they would appear to pass each other in the sky all the time.

However, the planets' orbits don't all exist perfectly in the same plane. They're all tilted a little, so that planets don't all fall exactly along a line in the sky. Sometimes a planet is a little above the plane, and sometimes a little below. It's extraordinarily rare for them to actually pass *directly* in front of each other. Usually they approach the same area of the sky, getting perhaps to within the width of the full Moon, then separate again. Often they never even get that close to each other, passing many degrees apart. For this reason, surprisingly, it's actually rather rare for more than two planets to be near each other in the sky at the same time.

Every so often, though, it does happen that the cosmic clock aligns a bit better than usual, and some of the major planets will appear to be in the same section of the sky. In 1962, for example, the Sun, the Moon, and all the planets except Uranus, Neptune, and Pluto appeared to be within 16 degrees of each other, which is roughly the amount of sky you can cover with your outstretched hand. Not only that, but there was also a solar eclipse, making this a truly spectacular event. The Moon and the Sun were as close as they could possibly be, since the Moon was directly in front of the Sun. In the year 1186, there was an even tighter alignment, and these planets could be contained with a circle just 11 degrees across.

On May 5, 2000, at 8:08 A.M. Greenwich Time, the planets Mercury, Venus, Mars, Jupiter, and Saturn were in very roughly the same section of the sky. Even the new Moon slid into this picture at that time, making this a very pretty family portrait indeed, although it was a bit of a dysfunctional family. This particular alignment wasn't a very good one, and even if it had been, the Sun was between us and the planets like an unwelcome relative standing in front of the TV set during the football playoffs.

The fact that this wasn't a particularly grand alignment is easy to show. The planets involved were within about 25 degrees of each other. That's half again as far as the 1962 alignment, and more than twice as bad as the one in 1186. Both of these years, it should be noted, are ones in which the Earth was *not* destroyed. As a matter of fact, there have been no fewer than 13 comparable alignments in the past millennium, and in none of them was there any effect on the Earth.

Still, this hardly even slowed the doomsayers down. The combined gravity of the planets, they claimed, was still enough to destroy the Earth. Since we're still here, we know that wasn't true. Still, it pays to look at this a little more carefully.

The force of gravity is overwhelming in our daily lives. It keeps us stuck to the Earth unless we use tremendously powerful rockets to overcome it. Gravity is what holds the Moon in orbit around the Earth, and the Earth around the Sun. It makes parts of us sag as we get older, and even manages to keep the highest vertical leap of the greatest basketball players in the world under a measly meter.

But gravity is also mysterious. We cannot see it, touch it, or taste it, and we know that the math involved in predicting it can be complicated. So it's easy—and all too human, I'm afraid—to assign all sorts of powers to gravity without really understanding it. In a way, understanding the effects of gravity is like a prize fight: science and its retinue of observations, facts, and math versus our superstitions, emotions, and the human power to jump to conclusions without much evidence. Which side will win the day?

Let's take a quick look at what we know about gravity: for one thing, it gets stronger with mass. The more massive an object is, the stronger its gravity. From a kilometer away, a mountain has more of a gravitational effect on you than, say, a Volkswagen.

However, we also know that gravity gets weaker very quickly with distance. That Volkswagen may be a lot smaller than the mountain, but its gravity will actually overwhelm the gravity of the mountain if the car is close and the mountain far away.

It's all relative. Indeed, the planets are massive. Jupiter tips the scales at over 300 times the Earth's mass. *But it's far away.* Very far. At its absolute closest, Jupiter is about 600 million kilometers (400 million miles) away. Even though it has 25,000 times the Moon's mass, it is nearly 1,600 times farther away. When you actually do the math, you find that the effect of Jupiter's gravity on the Earth is only about 1 percent of the Moon's!

Despite the old saying, size doesn't matter; distance does.

If you add up the gravity of *all* the planets, even assuming they are as close to the Earth as possible, you still don't get much. The Earth's tiny little Moon exerts 50 times more gravitational force on us than all the planets combined. The Moon is small, but it's close, so its gravity wins.

And that's true only if the planets are lined up as close to the Earth as they can get. As it happens, on May 5, 2000, the planets were on the *far* side of the Sun, meaning that you need to add the diameter of the Earth's orbit—another 300 million kilometers (185 million miles)—to their distances. When you do, the combined might of the planets is easily overwhelmed by the gravity of a person sitting next to you in that Volkswagen. Sorry, doomsayers, but round 1 of this fight goes to science.

Usually at this point I am challenged by some people who say that it isn't the gravity of the planets that can cause damage, it's the *tides.* Tides are related to gravity. They are caused by the *change* in gravity over distance. The Moon causes tides on the Earth because at any moment one side of the Earth is nearer the Moon than the other. The side nearer the Moon therefore feels a slightly higher gravitational force from the Moon. This acts to stretch the Earth a tiny amount. We see this effect as a raising and a lowering of the sea level twice a day, which is what most people normally think of as tides.

Earthquakes are caused by the movement of huge tectonic plates that make up the Earth's crust. They rub against each other, usually smoothly. However, sometimes they stick a bit, letting pressure build up. When enough pressure builds up, the plates slip suddenly, causing an earthquake. Since tides can stretch an object, it's reasonable to ask whether tides can trigger earthquakes. Are we still doomed?

Happily, no. When doomsayers bring up tides, they are shooting themselves in the foot. The force of tides fades even faster with distance than gravity. If the force of gravity on Earth is piddly for the planets, then tides are even weaker. Comparing the Moon, again, to all the combined might of the planets, we find that the Moon has *20,000 times* the tidal force of all the other planets in the solar system, even at their closest approach to the Earth. Remember, in May 2000 the planets were *as far away as they possibly could be*. The tidal force was so small that even the finest scientific instruments on the planet were not able to measure it. Round 2 of the fight goes to science as well.

Math and science show pretty definitively that the gravity and tides of the planets are too small to have any effect on the Earth. However, it would be foolhardy to assume that emotions are swayed by logic. In one sense, the side of science is lucky: since the planets were all on the far side of the Sun, we had to look past the Sun to see them. That means they were only up during the day, when they are practically invisible. It would not have helped the situation if people could actually look up at night and see the planets approaching each other, even if it were a pretty weak grouping.

Still, even armed with hard numbers, it's always an uphill fight to battle the doomsayers. There were a lot of people out there trying to make money by scaring people about the alignment. Certainly some of these folks were honest, if misguided. Richard Noone wrote a book about the alignment, *5/5/2000 Ice: The Ultimate Disaster*, where he claims that the Earth's axis would tilt due to the combined pull of the planets, plunging the Earth into an ice age. Noone was sincere, and felt he had done the research to back up his claims. The problem is, his research involved almost no astronomy at all. He related *Bible* prophesies, Nostradamus, and even the shape of the Great Pyramid in Egypt to a disaster in the year 2000, and figured the planets must have something to do with it. Yet, in his meandering book precious little space is devoted to the planets, and nowhere—nowhere!—does he talk about the actual measurable effects of the planets.

I am almost willing to give Noone the benefit of the doubt and assume he really was concerned about global catastrophe. But I wonder: if he really felt that the Earth would be destroyed on May 5, 2000, why not give away his book for free so that people could be warned? I can't imagine he thought the royalties he got on the book would be worth much on May 6.

Noone wasn't even the first. In the 1980s astronomer John Gribbin and his coauthor Stephen Plagemann wrote an infamous book entitled *The Jupiter Effect,* which claimed—again, without the benefit of any math—that the gravity of the planets would affect the Sun, causing more solar activity, causing a change in the Earth's rotation, causing massive earthquakes. This tissue-thin string of suppositions led them to predict very matter-of-factly that Los Angeles would be destroyed in 1982. The book was a runaway best seller.

When, in fact, L.A. was *not* destroyed as predicted, Gribbin and Plagemann wrote *another* book called *Beyond the Jupiter Effect,* making excuses about why things didn't work out quite as they had predicted, and of course they never simply admitted they were wrong. You may not be surprised to find out that this second book was another best seller. It's *possible,* barely, that the first book was a simple mistake and they honestly believed what they preached. The motivation for the second book perhaps isn't as clear.

If Noone and Gribbin were simply misguided, during the May 2000 alignment the "Survival Center" company was far more deliberate. Peddling disaster nonsense, this company had a web site promoting Noone's book as well as equipment to help you survive the oncoming onslaught. On their web site (http://www.zyz.com/survivalcenter/echange.html), they reported,

> Some scientists have already reported a distinct increased wobble to the earth as it begins to respond to the gravitational pull of the alignment . . . predictions [of the results of the alignment] range from a few earthquakes to major earth crust movement (slippage), polar ice cap movement, sea levels rising 100 to 300 feet or more, huge tidal waves, high winds 500 to 2000 miles per hour, earthquakes so massive that Richter 13 or more could be possible, both coasts of USA under water, magnetic shift and much more.

In 1998, I e-mailed them with this question: "May I ask, who are the people making these predictions? I would appreciate being able to contact them so that I may present my arguments on this issue." They replied, basically informing me that I had my sources and they had theirs. They wouldn't tell me who their sources or what their credentials were. I'm not surprised; backed up by hard science, no one can truthfully claim that the planets can have any sudden and catastrophic effect on the Earth. I would have serious doubts about the Survival Center's expertise in this matter anyway, even if they *had* revealed their sources. My opinion in situations such as these is, "Beware the science of someone trying to sell you something."

Of course, I'm trying to sell you something as well. But in my case, I'm peddling skepticism. You can go and find this stuff out for yourself if you try hard enough. The math isn't hard, and the conclusions are, well, *conclusive*.

My only real complaint about this whole alignment business—besides the vultures preying on people's fears—is that we weren't able to see it. The Sun was in the way, completely overwhelming the relatively feeble light from the planets and our Moon. So, not only were we denied the excitement of impending disaster, but also we couldn't even take a picture of it to show our grandkids! And we'll have to wait until September of 2040 for the next good alignment. At least that one *will* be visible at night.

15

🐛🐛🐛🐛🐛🐛

Meteors, Meteoroids, and Meteorites, Oh My!: The Impact of Meteors and Asteroids

On December 4, 2000, at roughly 5:00 P.M., something fell out of the sky and landed in David and Donna Ayoub's backyard in Salisbury, New Hampshire. Witnesses say the object was moving rapidly and glowing hot. When it landed, it set two small fires a couple of meters apart on the Ayoubs' property. The couple quickly ran outside to put them out.

The event certainly brought a lot of attention to the town. At first, it was a small story in the news section of the local newspaper, the *Concord Monitor*. However, the story was quickly picked up by an e-mail list sent out to astronomers interested in asteroid, meteor, and comet impacts. Soon the Ayoubs were receiving phone calls and were welcoming news media from all around the world. Everyone wanted to hear about what they saw, and most people assumed it was a meteorite impact.

I was skeptical when I heard the story the next day. I decided to look into this myself, so I phoned several of the witnesses. These people were sincere, and really wanted to know what had happened. After listening to them I believe that something truly did fall from the sky and set two fires. However, I don't think it was a meteorite, whatever it was.

Why don't I believe it was a meteorite? Well, that's a tale of bad astronomy.

※ ※ ※

I've always felt sorry for small meteors.

A given meteoroid may spend billions of years orbiting the Sun, perhaps first as part of a magnificent comet or an asteroid. Finally, after countless times around the Sun, its path intersects the Earth. It closes in on the Earth at a velocity that can be as high as 100 kilometers (60 miles) per second. Upon contact with our atmosphere, the tremendous speed is converted to heat, and, unless the meteoroid is too big (say, bigger than a breadbox), that heat vaporizes the tiny rock.

From our vantage point on the Earth's surface, the meteoroid generates a bright streak that may or may not be seen by human eyes. After all those billions of years, the life of that small rock is over in a few seconds, and no one might even see it.

But its story doesn't end there. When I am asked to name the most common example of bad astronomy, I almost always answer: meteors. Nearly everyone who is capable has seen a meteor flashing across the sky, yet, ironically, most people don't understand them at all.

Worse, even the naming of the phenomenon gets confused. Some people call them "shooting stars," but of course they aren't really stars. In chapter 3, "Idiom's Delight," I go over the three names describing the various stages of the rock: The solid part is called a **meteoroid** both while out in space and passing through our atmosphere, the glow of the meteoroid as it passes through the atmosphere is called a **meteor,** and it's a **meteorite** when (or if) it hits the ground.

But giving them names doesn't help much. We need to know what's going on during those stages.

A meteoroid starts out life as part of a bigger body, usually as either a comet or an asteroid. Asteroids can collide with each other, violently flinging out material or, in a worst-case scenario, shattering the parent body completely. Either way, you get debris going off rapidly in all directions. That debris can take on new orbits, where it might eventually cross paths with the Earth. When that happens, we might see a single bright meteor flash across the sky. Since the bits of meteoroid may be coming from any random direction in space, we see them come from any random point in the *sky*, traveling in a random direction. We call these **sporadic** meteors.

Cometary meteors are different. Comets are about the same size as asteroids but have a different composition. Instead of being mostly rock or metal, comets are more like frozen snowballs; rocks (from pebble size to kilometers across) held together by frozen material like water, ammonia, and other ices. When a comet gets near the Sun, the ice melts, and little bits of rock can work loose. This type of debris stays in roughly the same orbit as the comet for a long time. Not forever, though, because the orbit can be affected by the gravity of nearby planets, the solar wind, and even the pressure of light from the Sun. But the debris orbits are generally similar to that of the parent comet.

When the Earth plows through a ribbon of this meteoroidal debris, we see not one but many meteors. Usually it takes a few hours or nights to go all the way across the debris path, so we get what are called **meteor showers**—like a rain of meteors. We pass through the same debris ribbons every year at about the same time, so showers are predictable. For example, every year we pass through the orbital debris of the comet Swift-Tuttle, and we see a meteor shower that peaks around August 12 or 13.

Meteor showers create an odd effect. Imagine driving a car through a tunnel that has lights all around the inside. As you pass them, the lights all seem to be streaking outward from a point ahead of you in the tunnel. It's not real, since the lights are really all around you, but an effect of perspective. The same thing happens with meteor showers. The Earth's orbit intersects the meteoroid stream at a certain angle, and that doesn't change much from year to year. Like the lights in the tunnel, the meteors flash past you from all over the sky, but if you trace the path of every meteor backwards, they all point to one spot called the **radiant.** This point comes from a combination of the direction the Earth is headed in space and the motion of the meteoroids themselves. The radiant is almost literally the light at the end of the tunnel.

So the meteor shower I mentioned above not only recurs in time but in space, too. Every August those meteors appear, and they seem to flash out of the sky from the direction of the constellation Perseus. Showers are named after their radiant, so this one is called the Perseids.

One of the most famous showers comes from the direction of Leo every November. The Leonids are interesting for two reasons: One is that, relative to us, the parent comet orbits the Sun backwards. That means we slam into the meteoroid stream head-on. The meteoroids' velocity adds to ours, and we see the meteors flash across our sky particularly quickly.

The second interesting thing is that the meteoroid stream is clumpy. The comet undergoes bursts of activity every time it gets near the Sun (every 33 years or so), and this ejects lots of bits of debris. When we pass through these concentrated regions, we see not just dozens or hundreds of meteors an hour but sometimes thousands or even tens of thousands. This is called a **meteor storm.** The celebrated storm of 1966 had hundreds of thousands of meteors an hour, which means, had you been watching, you would have seen many meteors whizzing by *every second*. It must have really seemed as if the sky were falling.

So that's why we get meteors. But why are they so bright? Almost everyone thinks it's friction—our atmosphere heating them up, causing them to glow. Surprise! That answer is wrong.

When the meteoroid enters the upper reaches of the Earth's atmosphere, it compresses the air in front of it. When a gas is compressed it heats up, and the high speed—perhaps as high as 100 kilometers per second—of the meteoroid violently shocks the air in its path. The air is compressed so much that it gets really hot, hot enough to melt the meteoroid. The front side of the meteoroid— the side facing this blast of heated air—begins to melt. It releases different chemicals, and it's been found that some of these emit very bright light when heated. The meteoroid glows as its surface melts, and we see it on the ground as a luminous object flashing across the sky. The meteoroid is now glowing as a meteor.

Here I am guilty of a bit of bad astronomy myself. In the past, I've told people that friction with the air heats the meteoroid and, as I said above, this is the usual explanation given in books and on TV. However, it's wrong. In reality, there is actually very little friction between the meteoroid and the air. The highly heated, compressed air stays somewhat in front of the meteoroid, in what physicists call a **standoff shock.** This hot air stays far enough in front of the actual surface of the rock that there is a small pocket of relatively slow-moving air directly in contact with it. The heat from the compressed air melts the meteoroid, and the slow-moving air blows off the melted parts. This is called **ablation.** The ablated particles from the meteoroid fall behind, leaving a long glowing trail (sometimes called a train) that can be kilometers long and can stay glowing in the sky for several minutes.

All of these processes—the huge compression of air, the heating of the surface, and the ablation of the melted outer parts—happen very high in the atmosphere, at altitudes of tens of kilometers. The energy of the meteoroid's motion is quickly dissipated, slowing it down rapidly. The meteoroid slows to below the speed of sound, at which point the air in front is no longer greatly compressed and the meteor stops glowing. Regular friction takes over, slowing the meteoroid down to a few hundred kilometers per hour, which is really not much faster than a car might travel.

This means that it takes a few minutes for an average meteoroid to pass the rest of the way through the atmosphere to the ground. If it impacts the ground, it is called a meteorite.

This leads to yet another misconception about meteors. In practically every movie or television program I have ever seen, small meteorites hit the ground and start fires. But this isn't the way it really happens. Meteoroids spend most of their lives in deep space and are, therefore, very cold. They're only heated briefly when they pass through the atmosphere, and they're not heated long enough for that warmth to reach deep inside them, especially if they are made of rock, which is a pretty decent insulator.

In fact, the hottest parts ablate away, and the several minutes it takes for the meteoroid to get to the ground let the outer parts cool even more. Plus, it's traveling through the cold air a few kilometers off the ground. By the time it impacts, or shortly thereafter, the extremely frigid inner temperature of the meteoroid cools the outer parts very well. Not only do small meteorites *not* cause fires, but many are actually covered in *frost* when found!

Large meteorites are a different story. If it's big enough—like a kilometer or more across—the atmosphere doesn't slow it much. To really big ones, the atmosphere might as well not exist. They hit the ground at pretty much full speed, and their energy of motion is converted to heat. A *lot* of heat. Even a relatively small-ish asteroid a hundred meters or so across can cause widespread damage. In 1908, a rock about that size exploded in the air over a remote, swampy region in Siberia. The Tunguska Event, as it's now called, caused unimaginable disaster, knocking down trees for hundreds of kilometers and triggering seismographs across the planet. The event was even responsible for a bright glow in the sky visible at midnight in England, thousands of kilometers from the blast. The fires it started must have been staggering.

Understandably, such events are a cause of concern. Even little rocks—well, maybe the size of a football stadium—can have big consequences. But it does take a fair-sized rock to do that kind of damage. Little ones, and I mean really little ones, like the size of an apple or so, usually don't do more than put on a pretty show. I remember seeing a **bolide,** as the brightest meteors are called, as I walked home from a friend's house when I was a teenager. It lit up the sky, bright enough to cast shadows, and left a tremendous train behind it. I can still picture it clearly in my mind, all these years later. Sometime afterward I calculated that the meteoroid itself was probably not much bigger than a grapefruit or a small bowling ball.

But the big meteorites worry a lot of people, as well they should. Very few scientists now doubt that a large impact wiped out the dinosaurs, as well as most of the other species of animals and plants on the Earth. That impactor was probably something like 10 kilometers (6 miles) or so in diameter, and left a crater hundreds of kilometers across. The explosion may have released an unimaginable 400 million megatons of energy (compare that to the largest nuclear bomb ever built, which had a yield of about 100 megatons). It's no surprise that some astronomers stay up nights (literally) thinking about them.

There are teams of astronomers across the world looking for potential Earth impactors. They patiently scan the sky night after night, looking for the one faint blip that moves consistently from one image to the next. They plot the orbit, project it into the future, and see if our days are numbered.

No one has found such a rock yet. But there are a lot of rocks out there. . . .

Suppose that sometime in the near future the alarm is pulled. An asteroid as big as the Dinosaur Killer is spotted, and it will soon cross paths with us. What can we do?

Despite Hollywood's efforts, the answer is probably not to send a bunch of wisecracking oil riggers in souped-up rocket ships to the asteroid to blow it up at the last second. That may have worked in the 1998 blockbuster *Armageddon,* but in real life it wouldn't work. Even the largest bomb ever built would not disintegrate an asteroid "the size of Texas." (Not that *Armageddon* was terribly accurate in anything it showed; about the only thing it got right was that there is an asteroid in it, and asteroids do indeed exist.) In the same year, the movie *Deep Impact* depicted a comet getting shattered by a bomb shortly before it entered the Earth's atmosphere. That's even worse! Instead of a single impact yielding an explosion of billions of megatons, you'd get a billion impacts each exploding with a yield of many megatons. In his fascinating book, *Rain of Iron and Ice* (New York, Helix Books, 1996), University of Arizona planetologist John Lewis calculates that breaking up a moderately sized asteroid can actually increase the devastation by a factor of four to ten. You'd spread the disaster out over a much larger area of the Earth, causing more damage.

If we cannot blow it up, then what? Of course, the best option is for it to miss us in the first place, so we'd have to shove it aside. The orbit of an asteroid can be altered by applying a force to it. If enough time is available, like decades, the amount of force can be small. A larger force is needed if time is short.

There are several plans for pushing such rocks out of the way. One is to land rockets on the surface and erect a giant solar sail. The sail, made of very thin Mylar with an area of hundreds of square kilometers, would catch the solar wind and also react to the minute pressure of sunlight. It would impart a gentle but constant force, moving the rock into a safer trajectory.

Another plan is more blunt: attach rockets to the asteroid and use them to push it. This has the engineering difficulty of just how you'd strap boosters to a rock in the first place.

Ironically, Hollywood came close to another good plan. Instead of blowing the rock up, we use nuclear weapons to *heat* the asteroid. Again, in *Rain of Iron and Ice,* Lewis finds that a small nuclear explosion (he implies a yield of about 100 kilotons) would suffice. Exploded a few kilometers above the surface, the intense heat of the explosion would vaporize material off the surface of the asteroid. This material would expand outward, and, like a rocket, push the asteroid in the other direction. Lewis mentions that this has two benefits: it prevents the impact, and also removes a nuclear weapon from the Earth. This is the favored method of all the people who have studied it.

All of these methods have a subtle assumption attached, that we understand the structure of asteroids and comets. In reality, we don't. Asteroids come in many flavors; some are iron, some stony. Others appear to be no more than loose piles of rubble, barely held together by their own gravity. Without knowing even the most basic information about asteroids, we are literally shooting in the dark.

As with most problems, our best weapon is science itself. We need to study asteroids and comets, and study them up close, so that we can better understand how to divert them when the time comes. On February 14, 2000, the NASA probe Near Earth Asteroid Rendezvous entered orbit around the asteroid Eros. The amount learned from the mission is astounding, such as the surface structures and mineral composition of the asteroid. More probes are planned, some of which are ambitious enough to actually land on asteroids and determine their internal structure. We may yet learn how to handle dangerous ones when the time comes.

There is an interesting corollary to all this. If we can learn how to divert an asteroid instead of merely blowing it up, that means we can *steer* it. It may be possible to put a dangerous asteroid into a safe orbit around the Earth. From there we could actually set up mining operations. Based on spectroscopic observations of meteorites and asteroids, Lewis estimates that an asteroid 500 meters across would be worth about $4 *trillion* in cobalt, nickel, iron, and platinum. The metal is pure and in its raw form, making mining relatively easy, and the profit from such a venture would be more than enough to pay off any initial investment. And that's a *small* asteroid. Bigger ones abound.

Science fiction author Larry Niven once commented that the reason the dinosaurs became extinct is that they didn't have a space program. *We* do, and if we have enough ambition and enough reach, we can turn these potential weapons of extinction into a literal gold mine for humanity.

🐦 🐦 🐦

Until then, we don't have too many options. Maybe we can divert the big one when the time comes, but for now all we can do is imagine what an impact might be like. Unfortunately, movies have had their own impact. Anytime an unexplained phenomenon involves something falling from the sky, meteors are usually blamed.

Which brings us back to the Ayoubs, still searching for a meteorite in their backyard in Salisbury, New Hampshire. Initially, this night visitor sure did sound like the usual description of a meteorite. But my knowledge of their behavior was telling me otherwise. As I said, meteorites won't cause fires unless they are very big. But other things didn't add up, either. The path was described as an arc, while a meteor's trajectory would have been straight down. Also, no meteorite was ever found, despite a dedicated search. I mentioned to the property owner that meteorites can be sold for quite a bit of money, so he had strong incentive to find it. I never heard of anyone finding anything.

In the end, these events usually have some mundane, terrestrial cause. I would bet money that it was someone setting off fireworks in the thick woods near the Ayoubs' house. This is a guess on my part, and it may be wrong. We may never know what started those fires, but we know what it *wasn't*. We can blame Hollywood for our mistaken understanding of meteorites, but we can't blame everything else on the poor things themselves.

16

ขขขขขข

When the Universe
Throws You a Curve:
Misunderstanding the
Beginning of It All

Astronomy sometimes has a way of making people feel small. For most of our history we humans have been pretty self-important. We believe that the gods pay special attention to us, even intervening in our daily affairs. We claim territory for ourselves, and ignore what goes on outside those borders. Why, we've even said the whole universe revolves around us!

But the universe is under no obligation to listen to our petty boasts. Not only are we not at the center, but also there really isn't a center at all. To see why, we need to look into the past a bit, back into our own history.

For thousands of years it was thought that the Earth was the center of the universe and the heavens spun around us. Certainly, observations support that belief. If you go outside and look up for even a few minutes, you'll see that the whole sky is moving. But you don't feel any movement, so clearly the Earth is fixed, and the sky moves.

Even today, when we know better, we still talk as though this is the way things are: our vocabulary reflects the geocentric universe. "The Sun rose at 6:30 this morning" is less accurate than saying, "From my fixed location on the surface of the spherical Earth, the horizon moved below the apparent position of the Sun at 6:30 this morning." But it *is* easier to say.

This Earth-centered model was fine-tuned by the Greek astronomer Ptolemy around A.D. 150 or so. People used it to predict planet positions, but the planets stubbornly refused to follow the model. The model was "refined"—that is, made more complicated—but it never quite made the grade.

Eventually, a series of discoveries over the centuries removed the Earth from the center of the universe. First, Nicolaus Copernicus presented a model of the solar system in which the Earth went around the Sun, rather than vice-versa. His model wasn't really all that much better than Ptolemy's model at figuring out where the planets would be. But then Johannes Kepler came along a few centuries later and tweaked the model, discovering that the planets orbit in ellipses instead of circles, and things were a lot better.

So with Copernicus's model it looked like the Sun was the center of the universe. That's not as good as having the Earth there, but it's not too bad.

Then around the turn of the twentieth century, Jacobus Kapteyn tried to figure out how big the universe was. He did this in a simple way: he counted stars. He assumed that the universe had some sort of shape, and that it was evenly distributed with stars. If you saw more stars in one direction, then the universe stretched farther that way.

He found an amazing thing: the Sun really *was* the center of the universe! When he mapped out the stars, the universe was blobby, like an amoeba, but it seemed to be fairly well centered on the Sun. Maybe the ancients were right after all.

Or not. What Kapteyn didn't realize is that space is filled with gas and dust, which obscures our view. Imagine standing in the middle of a vast, smoke-filled room, like an airplane hangar. You can only see, say, 20 meters in any direction because smoke blocks your vision. You have no idea what shape the room is; it might be a circle, or a square, or a pentagram. You don't even know how big it is! It could have walls just a meter beyond your vision, or it could stretch halfway to the Moon. You can't tell just by looking.

But no matter how big or what shape, the room will look like it is about 20 meters in radius and centered smack dab on you.

That was Kapteyn's problem. Because he could only see out to a few hundred light-years before gas and dust blocked his view, he thought the Milky Way, which was then considered to be the whole universe, centered on us. However, observations by another astronomer, Harlow Shapley, in 1917 revealed that we are not at the center of the Milky Way, but indeed displaced quite a bit from the center.

Do you see the pattern? First the Earth was the center of every-thing—hurrah! Then, well, ahem. Maybe the Sun still is—yay! But then, yikes, actually we're way out in the suburbs of the Galaxy. Well, this was getting downright insulting.

But the worst humiliation was yet to come. Kapteyn's universe, as it was called, was about to collapse. Or, more aptly, explode.

Observations by Edwin Hubble, after whom the space telescope is named, showed that our Milky Way galaxy was just one of thou-sands and perhaps millions of other galaxies. What was thought to be the whole universe was really only just a single island of stars floating in space. Instead of being at the center of everything, we were just another face in the crowd.

When Hubble analyzed the light given off by these other galax-ies, he got what may be considered the single biggest surprise ever sprung on a scientist. He found that almost all these myriad galax-ies were rushing away from us. It was as if we were a cosmic pariah, and everything else in the universe was falling all over itself trying to get away from us.

Make no mistake: this is really weird, and completely unex-pected. The universe was thought to be static, unchanging. Yet Hubble found that it's on the move. It's hard to underestimate the impact of these observations. And there was more: Hubble found that not only were galaxies all rushing away from us, but also the ones farther away were moving faster than the ones near us. The tools of the time didn't let him look at galaxies that were terribly far away, but more recently, as bigger and more sensitive telescopes have come on line, we have found that Hubble was right. The far-ther away a galaxy is, the faster it appears to recede from us.

It didn't take long for people to realize that this was characteristic of an explosion. If you blow up a bomb, then take a snapshot of the explosion a few seconds later, you see how shrapnel that's farther from the center must be moving faster. The fastest bits move the most in a given time, while slower bits haven't moved out as far.

This implies that the universe started in a gigantic explosion. You can think of it this way: if all the galaxies are moving away from us as time goes on, then they must have been closer in the past. If you reverse time's arrow and let it run backwards, there must have been a time in the past when everything in the universe was crushed into a single point. Let time run forward again, and BANG! everything is set in motion.

And what a big bang it was, starting up the universe and sending it flying. Could this be right? Did the universe start out as a single point that exploded outwards? Perhaps no single scientific theory has stirred people, incited their anger, their confusion and, indeed, their awe more than the Big Bang theory. I suspect that even Darwin's observations on evolution may have to take back seat to the biggest bang of them all.

But it does have one comforting aspect: it says we are at the center, because everything is rushing away from us . . .

. . . or does it? Let's use an analogy. Imagine you are sitting in a movie theater, and the seats are packed together so closely that they are touching. Furthermore, the seats are all on movable tracks. I hit a button, and suddenly every seat moves so that there is now one meter separating each chair. Your nearest neighbors are all one meter away, in front of you, behind you, on your left, and on your right. The next seats over are all *two* meters away, and the next ones from those are *three* meters away, and so on. But wait! That's true for any seat in the house. If you got up and moved into a seat a couple of rows up, and we repeated this experiment, you would see exactly the same thing. The next seats over would be one meter away, and the ones past that would be two meters away, and so forth.

So no matter where you sit, it looks like all the seats are rushing away from you. It doesn't matter if you are actually in the center seat or not!

Also, the seats farthest away from you appear to be moving the fastest. The seats next to you moved one meter when I hit the button, but the next ones moved two meters, and so on. Again, no matter where you sit, you'd see the same thing: it looks like all seats are moving away, and that the ones that are farther away move the fastest.

That is *exactly* what Hubble found. Shakespeare said, "All the world's a stage," not realizing that, in a way, all the universe is a movie theater. Scientists studying Hubble's observations quickly realized that the universal expansion may be real, but it gives the illusion that we are at the center, when we may not be at the center at all.

And if that's not weird enough, the universe still has some tricks up its sleeve.

With stuff this bizarre going on, it's no surprise to find Einstein lurking somewhere nearby. Einstein was busily pondering the universe in the years before Hubble's shocking discoveries. He was applying some pretty hairy math to the problem, and came across a difficulty. The universe, he discovered, should not be here. Or, more precisely, that something was supporting it against its own gravity. Left to itself, the universe's gravity would cause all the galaxies to attract each other, and the universe would quickly collapse like a soufflé after the oven door is slammed. Before Hubble, remember, it was thought that the universe was unchanging. Something must be counteracting gravity, so Einstein decided to add a constant to his equation that would be a sort of antigravity. He didn't know what it was, exactly, but he figured it had to be there.

Or so he thought. When he found out along with the rest of the world that the universe was expanding, he realized that the expansion itself would counteract gravity, and he didn't need his cosmological constant. He discarded it, calling it "the biggest blunder of my life."

It's too bad, really. As astronomer Bob Kirshner once pointed out to me, given what Einstein knew at the time, he could have actually *predicted* the expansion of the universe. Why, he'd have been famous!

Anyway, what Einstein came to understand in later years is that the universe is a peculiar place. First, he realized that space is a *thing*. What that means is, it was always thought that space was just a place in which stuff existed, but space had no real presence itself. It was just *space*. But Einstein saw that space was a tangible thing, like a fabric into which the universe was woven. Gravity could distort that fabric, bending space itself. A massive object like a planet or a star (or, on a smaller but no less real a scale, a lobster or a toothbrush or a nail) warps space.

A common analogy compares our three-dimensional space to a two-dimensional rubber sheet. Stretched out, that sheet represents space. If you roll a tennis ball across it, the ball will move in a straight line. But if you put, say, a bowling ball on it, the sheet will get a funnel-shaped depression. If you then roll the tennis ball near the bowling ball, the path of the tennis ball will bend, curving around the bowling ball. That's what happens in the real universe: a massive object warps space, and the path of an object will bend when it gets near it. That warping is what we call gravity.

If space is itself a thing, then it's possible for space to have a shape. Indeed, the mathematics of cosmology strongly imply that space has some sort of shape to it. It's hard for us mere humans to wrap our brains around such a concept, so once again the two-dimensional analogy is pretty useful.

Imagine you are an ant, and you live on a flat sheet that extends infinitely in every direction. To you, there is no up or down; all there is is forward, back, left, and right. If you start walking, you can walk forever and always get farther from where you started.

But now I'm going to play a trick on you. I take you off the sheet and put you on a basketball. You can still only move in back or forth, ahead or back. But now, if you start walking straight, eventually you'll get back to where you started. Surprise! If you have a good grasp of geometry, you might realize that maybe your two-dimensional space is only a part of another, higher dimension. Furthermore, you can guess a bit about the shape of your space because your walk returned you to your starting point. That kind of space is *closed,* because it curves back onto itself. There is a boundary to it; it's finite.

Open space would be one that curves the other way, away from itself, so it takes on a saddle shape. If you lived in open space, you could walk forever and never get back to where you started.

These three spaces—open, closed, and flat—have different properties. For example, if you remember your high school geometry, you'll recall that if you measure the three inside angles of a triangle and added them together, you get 180 degrees. But that's only if space is flat, like a page in this book. If you draw a triangle on the surface of a sphere and do the same thing, you'll see that the angles always add up to more than 180 degrees!

Imagine: take a globe. Start at the north pole and draw a line straight down to the equator through Greenwich, England. Then go due west a quarter of the way around the globe. Now draw another line back up to the north pole. You've drawn a triangle, but each inside angle is 90 degrees, which adds up to 270 degrees, despite what your geometry teacher taught you. Actually, your teacher was just sticking with flat space; closed and open space can be quite different. In open space, the angles add up to *less* than 180 degrees.

So that ant, if it were smart enough, could actually try to figure out if its space is open, closed, or flat just by drawing triangles and carefully measuring their angles.

This is all well and good if you're an ant, but what about us, in our three-dimensional space? Actually, the same principles apply. Since space itself is warped, it can take on one of these three shapes, also called geometries. And, just like the ant, you could try taking a walk to see if you come back to where you started. The problem is that space is awfully big, and even the fastest rocket we can imagine would take billions or even trillions of years to come back. Who has that kind of time?

There's an easier way. Karl Friedrich Gauss was a nineteenth-century mathematician who worked out a lot of the math of the geometry of the universe. He actually tried to measure big triangles from three hilltops, but was unable to tell if the angles added up to more or less than 180 degrees.

There are still other ways. One is to look at incredibly distant objects and carefully observe their behavior. Using complicated physics, it's possible to determine the universe's geometry. At the moment, our best measurements show that the universe is flat. If it curves at all on large scales, it's very difficult to see.

Now let's imagine again that you're an ant, back on the ball. As a fairly smart ant, you might ask yourself: If my universe is curved, where is the center? Can I go there and look at it?

The answer is no! Remember, you're stuck on the *surface* of the ball, with no real concept of up or down. The center of the ball isn't on the surface, it's *inside,* removed into the third dimension, which you cannot access. You can search all you want, but you'll never find the center, because it's not in the universe as you know it.

The same can be said for own 3-D universe. If it has a center, it might not be in our universe at all, but in some higher dimension.

As it happens, even *this* might not be the case. Gauss showed mathematically that, as bizarre as it sounds, the universe can be curved without curving *into* anything. It just exists, and it's curved, and that's that. So it's not that we are curved into the fourth dimension, if there is such a thing. The fourth dimension may not exist at all, and our universe simply may not have a center.

This is the worse humiliation of all. To be removed from the center of the universe is one thing, and it's another to have it appear that we are at the center, only to realize that anywhere in the universe can make that claim. But then, to be told *there isn't any center at all* is the ultimate insult.

Maybe in a way it's the perfect equalizer. If we can't occupy the center of everything, at least no one else can, either.

❧ ❧ ❧

And yet we are still not done.

Einstein was just getting started when he realized that space was a tangible thing. Time, he found, was a quantity that in many ways was like space. In fact, space and time were so intertwined that the term **space-time continuum** was coined to describe the union.

He also realized that the moment of creation, the Big Bang, was more than just a simple (though all-encompassing) explosion. It was not an explosion *in* space, it was an explosion *of* space. Everything was created in the initial event, including space and time. So asking what there was before the Big Bang really has no meaning. It's like asking, where was I before I was born? You were *nowhere*. You didn't exist.

But time was created in the event as well. So asking what happened before the Big Bang is what we call an ill-posed question, another question with no meaning. The physicist Stephen Hawking likens it to asking, "What's north of the north pole?" Nothing is! The question doesn't even make sense.

We want it to make sense, because we are used to things happening in a sequence. I get up in the morning, I ride my bike to work, I make my coffee. What did I do before I woke up? I was sleeping. Before that? I got into bed, and so on. But face it, at some point there was a first event. In my case, it was a moment in January 1964, which probably happened because it was a cold night and my future parents decided to snuggle a bit.

But there was something even before that, and before *that*. Eventually, we run out of *thats*. There was a first moment, a first event. The Big Bang.

In television documentaries it's very common to show an animation of the Big Bang as an explosion, a spherical fireball expanding into blackness. But that's wrong! Since the explosion was the initial expansion of space itself, there isn't anything for the universe to expand into. The universe *is all there is*. There is no outside, any more than there was a time before the Big Bang. What's north of the north pole?

The illusion of living in a big expanding ball persists. I have a hard time shaking it myself. You would think that there was some direction to the center of the universe, and if you looked that way you'd see it. The problem is, the explosion is all around us. We are part of it, so it's everywhere we look: the biggest movie theater of them all.

Still confused? That's okay. I sometimes think even cosmologists get headaches trying to picture the fourth dimension and the curvature of space, though they'd never admit it. There's an expression in astronomy: cosmologists are often wrong, but never unsure of themselves.

Yet we continue to try to understand this vast universe of ours. Maybe Albert himself put it best: "The most astonishing thing about the universe is that we can understand it at all."

I cannot leave this topic without one final note. Historians studying medieval astronomy are beginning to come to the conclusion that, to the medieval astronomers, being at the center of the universe was not all that privileged a position to occupy. It was thought that all the detritus and other, um, *waste products* of the heavens fell to the center, making up the Earth. So instead of being an exalted position, the center of the universe was actually a rather filthy place to be. In the end, maybe not even having a center is better than the alternative.

PART IV

Artificial Intelligence

People believe weird things.

There are people who believe the Earth is 6,000 years old. Some people believe that others can talk to the dead, that a horoscope can accurately guide your day, and that aliens are abducting as many as 800,000 people a year.

I believe weird things, too. I believe that a star can collapse, disappearing from the universe altogether. I believe that the universe itself started as a Big Bang, possibly as a leak in space and time from another, older universe. I believe that there is a vast reservoir of hundred-kilometer-wide chunks of ice hundreds of billions of kilometers out from the Sun, yet I have never seen one of these chunks *in situ,* nor has any other person on Earth.

So, what's the difference? Why do I think it's wrong to believe that the Earth is young when I believe in things I've never seen?

It's because I have *evidence* for my beliefs. I can point to well-documented, rational, reproducible observations and experiments that bolster my confidence in my conclusions. The examples in the second paragraph above are not similarly supported. The people who believe in such things will bring out piles of evidence, certainly, but it's written on tissue paper. A solid cross-examination of the evidence finds it flimsy, fragile, and sometimes even fabricated. The experiments rely on hearsay, or secondhand information, or bad statistics, or a nonreproducible event. Such evidence does not ably support a belief system. And it's definitely not science.

This section has several chapters that deal with *pseudo*science, ideas that sound superficially like science, but aren't anything of the sort. The difference between science and pseudoscience is that science is repeatable, and makes specific predictions that can be tested, while pseudoscience generally relies on single, unrepeated events or predictions that are impossible to test. Of all the forms of bad astronomy, pseudoscience is the most pernicious. You might laugh at some of the attitudes presented; how could a modern person believe that NASA never sent men to the moon? Why would someone think that a fuzzy photo of a piece of ice floating outside a Space Shuttle window is evidence of an alien war with humans?

Odds are that you believe NASA sent men to the Moon. So why devote a whole chapter to the minority that doesn't? There are several reasons. The most important is to simply provide a rational and reasoned voice when such a voice is hard to find. People who promote pseudoscience sometimes use astronomy, twisting it beyond recognition, and it can be difficult even for astronomers to understand where the arguments go wrong, let alone someone who is not educated in astronomy.

Also, without an opposing voice, a given hoax (and other matters of pseudoscience) can become endemic. Sure, the true believers will never listen to someone like me, but for every one true believer there are perhaps ten others who want to know the truth—call them passive believers—but who are only hearing one side of the story. They need to hear the other side, science's side, and that's what I present here.

I receive letters all the time from people who initially believed or at least questioned the claims of a pseudoscientist, but upon reading a rational rebuttal realized the pseudoscientist was wrong. I have hope that rational thinking will win in the end, largely because science produces reliable results. Carl Sagan put it best: "Science is a way to not fool ourselves."

So let's take a look at who's fooling whom.

17

Appalled at Apollo: Uncovering the Moon-Landing Hoax

I t's an engaging story, and almost plausible.

NASA is in trouble. Contractors on the upcoming space mission were negligent, and made a mistake on one of the parts they were building. The mistake was discovered too late, and the part is already integrated with the rocket. They know the part will fail, ending the mission in catastrophe, so they tell NASA. However, NASA officials are under intense public pressure for a successful launch. They know that if they admit there is a problem, the space program (and therefore their paychecks) will grind to a halt. So they decide to launch anyway, knowing the mission will fail.

But the rocket they launch is a dummy, with no one on board. The real astronauts are spirited away to the Nevada desert to a hastily assembled movie set. Under physical threat, the astronauts are forced to obey the NASA officials, faking the entire mission. What they don't know is that NASA plans on murdering them to protect the secret, then claim that astronaut error killed them upon reentry. NASA officials would take a hit but eventually would be exonerated.

Does this scenario sound believable? It does to some people. The story certainly interested Warner Brothers, which made this script into the movie *Capricorn One* in 1978. It's a pretty good movie, actually, and stars the unlikely group of Eliot Gould, James Brolin, and none other than O. J. Simpson. But remember, it's just a movie. It's not real.

Or was it? Despite what the vast majority of the human population believes, some hold that the movie was portraying reality. NASA faked the whole Apollo Moon project, they claim, and instead of its being the most incredible technical achievement of all time, it is actually the greatest fraud perpetrated on mankind. They believe the fraud continues today.

Surprisingly, there appears to be a market for such a belief. James Oberg, an expert on space travel and its history, estimates that there may be 10 to 25 million people in the United States alone who at least have doubts that NASA sent men to the Moon. This number may be about right—a 1999 Gallup poll found that 6 percent of Americans, or about 12 million people, believe the NASA conspiracy theory, the same number found in a 1995 *Time/CNN* poll. Executives at the Fox Television network thought enough people would be interested in this idea that in 2001 they aired a one-hour program about NASA covering up a faked Moon landing. The program was aired twice in the United States, in February and again in March of 2001 (it was later broadcast in several other countries as well). Combined, the show had about 15 million viewers in the United States alone. Judging from the discussion groups on the web, the radio and television activity about it, and the vast number of e-mails I received in the following months, *something* about that program touched a nerve in a lot of people.

That such a huge number of people could seriously believe the Moon landings were faked by a NASA conspiracy raises interesting questions—maybe more about *how* people think than anything about the Moon landings themselves. But still, the most obvious question is the matter of evidence. What manner of data could possibly convince someone that the Moon still lays untouched by human hands?

The answer is in the photographs taken by the astronauts themselves. If you look carefully at the images, the hoax believers say, you'll see through the big lie.

My question is, whose big lie? The hoax-believers may not be lying, that is, prevaricating consciously and with forethought, but they're *certainly* wrong. Most don't think they are wrong, of course, and they sure like to talk about it. A web search using the words "Apollo Moon hoax" netted nearly 700 web sites. There are several books and even videos available, adamantly claiming that no man has ever set foot on the Moon.

The most vociferous of the hoax-believers is a man by the name of Bill Kaysing. He has written a book, self-published, called *We Never Went to the Moon* that details his findings about a purported NASA hoax. Most of his arguments are relatively straightforward. His "evidence" has been picked up by web sites and other conspiracy theorists and usually simply parroted by them.

The evidence worth considering usually comes in the form of pictures taken by the astronauts themselves either on the Moon or in orbit above it. Thousands of pictures were taken by the astronauts, and many of them are quite famous. Some made rather popular posters, and others have been seen countless times as part of news reviews on TV and in newspapers. The overwhelming majority were relegated to an archive where specialists interested in the lunar surface could find them. Most of these consist of picture after picture of the astronauts performing their duties on the surface, and they are unremarkable except for the fact that they show spacesuited human beings standing for the first time in history on the airless plain of an alien world.

Unremarkable, of course, unless you are looking for a dark undercurrent of a NASA conspiracy.

There are five basic concerns raised by the hoax-believers. These are: (1) there are no stars in the astronaut photos, (2) the astronauts could not have survived the radiation during the trip, (3) there is dust under the lunar lander, (4) the incredibly high temperature of the Moon should have killed the astronauts, and (5) the play of light and shadows in the surface indicates that the photos are faked. There are a host of other "problems," a few of which we'll look at after the main points, but let's look at the biggest first.

In the pictures taken by the Apollo astronauts, no stars can be seen. Far from being evidence of a hoax, this is evidence men *did* go to the Moon. The bright surface and highly reflective spacesuits meant short exposure times were needed to take properly exposed pictures, and the faint stars were too underexposed to be seen.

1. *No stars in the astronauts' photos*

A typical Apollo photograph shows a gray-white lunar land-scape, an astronaut in a blindingly white spacesuit performing some arcane function, a jet-black featureless sky, and sometimes a piece of equipment sitting on the surface, doing whatever it is it was built to do.

The hoax-believers put their biggest stake in these very pictures. Almost without exception, the first and biggest claim of the conspiracy theorists is that those pictures should show thousands of stars, yet none is seen! Kaysing himself has used this argument numerous times in interviews. On the airless surface of the Moon, the conspiracy theorists say, the sky is black, and therefore stars should be plentiful (see chapter 4, "Blue Skies Smiling at Me," for more about this phenomenon). The fact that they are *not* there, they continue, proves conclusively that NASA faked the images.

Admittedly, this argument is compelling. It sounds convincing, and it appeals to our common sense. When the sky is black at night here on Earth we easily see stars. Why should it not be true on the Moon as well?

Actually, the answer is painfully simple. The stars are too faint to be seen in the images.

During the day, the sky here on Earth is bright and blue because molecules of nitrogen in the air scatter the sunlight everywhere, like pinballs in a celestial pachinko game. By the time that sunlight reaches the ground, it has been bounced every which-way. What that means to us on the ground is that it looks like the light is coming from every direction of the sky and the sky appears bright. At night, after the Sun goes down, the sky is no longer illuminated and appears black. The fainter sky means we can see the stars.

On the Moon, though, there's no air, and even the daytime sky appears black. That's because without air, the incoming sunlight isn't scattered and heads right at you from the Sun. Any random patch of sky is *not* being illuminated by the Sun, and so it looks black.

Now imagine you are on the Moon, and you want to take a picture of your fellow astronaut. It's daytime, so the Sun is up, even though the sky is black. The other astronaut is in his white spacesuit, cavorting about in that bright sunlight, on that brightly lit moonscape. Here's the critical part: when you choose an exposure time for the camera, you would set the camera for a brightly daylit scene. The exposure time would therefore be very short, lest you overexpose the astronaut and the moonscape. When the picture comes out, the astronaut and the moonscape will be exposed correctly and, of course, the sky will look black. But you won't see any stars in the sky. The stars are there, but in such a short exposure they don't have time to be recorded on the film. To actually see stars in those pictures would require long exposures, which would utterly overexpose everything else in the frame.

Put it another way: if you were to go outside at night here on Earth (where the sky is still black) and take a picture with *exactly the same settings* that the astronauts used on the Moon, you would still see no stars. They are too faint to get exposed properly.

Some people claim that this still won't work because actually the Earth's air absorbs starlight, making them fainter, so stars should look brighter from the surface of the Moon. That's not correct; it's a myth that air absorbs a lot of starlight. Actually, our atmosphere is amazingly transparent to the light we see with our eyes, and it lets almost all the visible light through. I chatted with two-time Space Shuttle astronaut and professional astronomer Ron Parise about this. I asked him if he sees more stars when he's in space, and he told me that he could barely see them at all. He had to turn off all the lights inside the Shuttle to even glimpse the stars, and even then the red lights from the control panels reflected in the glass, making viewing the stars difficult. Being outside the Earth's atmosphere doesn't make the stars appear any brighter at all.

The accusation made by the hoax-believers about stars in the Apollo photographs at first may sound pretty damning, but in reality it has a very simple explanation. If the believers had asked any professional photographer or, better yet, any of the hundreds of thousands of amateur astronomers in the world, they would have received the explanation easily and simply. They also could easily prove it for themselves with a camera.

I am frankly *amazed* that conspiracy theorists would put this bit of silliness forward as evidence at all, let alone make it their biggest point. In reality, it's the *easiest* of their arguments to prove wrong. Yet they still cling to it.

2. *Surviving the Radiation of Space*

In 1958 the United States launched a satellite named Explorer 1. Among its many discoveries, it found that there was a zone of intense radiation above the Earth, starting at about 600 kilometers (375 miles) above the surface. University of Iowa physicist James Van Allen was the first to correctly interpret this radiation: it was composed of particles from the Sun's solar wind trapped in the Earth's magnetic field. Like a bar magnet attracting iron filings, the Earth's magnetic field captures these energetic protons and electrons from the Sun's wind, keeping them confined to a doughnut-shaped series of belts ranging as high as 65,000 kilometers (40,000 miles) above the Earth. These zones of radiation were subsequently named the **Van Allen belts**.

These belts posed a problem. The radiation in them was pretty fierce and could damage scientific instruments placed in orbit. Worse, the radiation could seriously harm any humans in space as well.

Any electronics placed on board satellites or probes need to be "hardened" against this radiation. The delicate and sophisticated computer parts must be able to withstand this bath of radiation or they are rendered useless almost instantly, fried beyond repair. This is an expensive and difficult process. It surprises most people to learn that the typical computer in space is as much as a decade behind the technology you can buy in a local store. That's because of the lengthy process involved in radiation-hardening equipment. Your home computer may be faster than the one on board the Hubble Space Telescope, but it would last perhaps 15 seconds in space before turning into a heap of useless metal.

Shuttle astronauts stay below the Van Allen belts, and so they do not get a lethal dose of radiation. The doses they do get are elevated compared to staying on the ground, to be sure, but staying below the belts greatly reduces their exposure.

Hoax-believers point to the Van Allen radiation belts as a second line of evidence. No human could possibly go into that bath of lethal radiation and live to tell the tale, they claim. The Moon landings must have been faked.

We've seen once before that basic logic is not exactly the hoax-believers' strong suit. It's not surprising they're way off base here, too.

For one, they are vastly confused about the belts. They claim that the belts "protect" the Earth from radiation, trapping it high above us. Outside the belts, they go on, the radiation would kill a human quickly.

That's not true, at least not totally. There are actually two radiation belts, an inner one and an outer one, both shaped like doughnuts. The inner one is smaller, and has more intense—and therefore more dangerous—radiation. The outer one is bigger but has less dangerous properties. Both belts trap particles from the solar wind, so the radiation is worst when an astronaut is actually *inside* the belts. I talked with Professor Van Allen about this, and he told me that the engineers at NASA were indeed concerned about the radiation in the belts. To minimize the risk, they put the Apollo spacecraft along a trajectory that only nicked the very inside of the inner belt, exposing the astronauts to as little dangerous radiation as possible. They spent more time in the outer belts, but there the radiation level isn't as high. The metal walls of the spacecraft protected the astronauts from the worst of it. Also, contrary to popular belief, you don't need lead shielding to protect yourself from radiation. There are different kinds of radiation; alpha particles, for example, are really just fast-moving helium nuclei that can be stopped by normal window glass.

Once outside the van Allen belts—contrary to the claims of the hoax-believers—radiation levels drop, so the astronauts were able to survive the rest of the way to the Moon. From the belts on out they were in a slightly elevated but perfectly safe radiation environment.

There was risk, though. Under normal circumstances, the solar wind is a gentle stream of particles from the Sun. However, there was a very real danger from solar flares. When the Sun's surface flares, there can be a dramatic increase in the amount of radiation the sun emits. A good-sized flare could indeed kill an astronaut, very nastily and gruesomely. In that sense, the astronauts were truly risking their lives to go to the Moon because solar flares are not predictable. Had there been a good flare, they might have died, farther from home than anyone else in history. Luckily, the Sun's activity was low during the missions and the astronauts were safe.

In the end, over the course of their trip to the Moon and back, the astronauts got, on average, less than 1 rem of radiation, which is about the same amount of radiation a person living at sea level accumulates in three years. Over a very long time that level of exposure might indeed be dangerous, but the round-trip to the Moon was only a few days long. Since there weren't any flares from the Sun, the astronauts' exposure to radiation was actually within reasonable limits.

Conspiracy theorists also argue that the radiation should have fogged the film used on the lunar missions. However, the film was kept in metal canisters, which again protected it from radiation. Ironically, modern digital cameras no longer use film; they use solid-state electronic detectors, which are sensitive to light. Like any other kind of computer hardware, these detectors are also very sensitive to radiation, and would have been next to useless on the Moon, even if they had been encased in metal. In that case, the older technology actually did a better job than would modern technology.

3. Dust on the Moon's Surface

The surface of the Moon is dusty. Before any machines landed on the Moon, no one really knew what the actual surface was like. Scientific analysis showed that the Moon's surface was rocky, and we could even determine the composition of some of the rocks. However, the actual texture of the surface was unknown. It was conjectured by some that the intense sunlight, consisting of ultraviolet light unfiltered by an atmosphere, might break down the rocks into a dust. Micrometeorite hits might do the same. But no one knew for sure if the dust even existed, or how deep it might be.

When the first soft landings were made by Soviet and American probes, it was found that the dust was only a few millimeters to centimeters thick. That was a great relief. No one wanted the Apollo astronauts to sink into a sand trap.

The dust on the Moon is peculiar. It is extraordinarily fine, like well-ground flour. It is also extremely dry, like everything else on the Moon. Unlike the Earth, the Moon has virtually no water at all anywhere on the surface.

Misunderstanding the properties of this dust in an airless environment leads to the breakdown of the next hoax-believer claim, dealing with the landing of the lunar module (or LM), the odd-looking contraption used by the Apollo astronauts to land on the Moon. The LM had four landing legs with disk-shaped feet at the ends, and between them was a powerful rocket used to slow the descent speed as the LM approached the surface.

The conspiracy-theorists claim that the rocket had a thrust of 10,000 pounds, and therefore should have left a substantial crater on the Moon's surface. Also, that much thrust would blow away all the dust underneath it. How could the lander's legs and the astronauts' boots leave imprints in dust? All that dust should be gone!

Both of these claims are wrong. First, the engine was capable of 10,000 pounds of thrust *at maximum*, but it wasn't simply a roman candle that burns at full thrust when lit. The engine had a throttle, basically a gas pedal, which could change the *amount* of thrust generated by the engine. When high over the surface of the Moon, the astronaut flying the lander would throttle the engine for maximum thrust, slowing the descent quickly. However, as the lander slowed, less thrust was needed to support it, so the astronaut would throttle back. By the time the lander touched down, the astronauts had cut the thrust to about 30 percent of maximum, just enough to compensate for the lander's own weight on the Moon.

Three thousand pounds of thrust still might sound like a lot, but the engine nozzle of the lander was pretty big. The bell was about 54 inches across, giving it an area of about 2,300 square inches. That 3,000 pounds of thrust was spread out over that area, generating a pressure of only about 1.5 pounds per square inch, which is really pretty gentle, less than the pressure of the astronauts' boots in the dust. That's why there is no blast crater under the lander; the pressure was too low to carve out a hole.

The second claim about dust near the lander is interesting. Why was there dust so close to the center of the landing site that both the lander legs and the astronauts' movements left tracks? This defies common sense, which says the dust should have all been blown away. However, our common sense is based on our experience here on Earth, and it pays to remember that the Moon is *not* the Earth.

Once again, we have to understand that the Moon has no air. Imagine taking a bag of flour and emptying it on your kitchen floor (kids: ask your parents first). Now stand over the flour, stick your face an inch or two above it, and blow as hard as you can.

When you stop coughing and sneezing from having flour blown into your nose, take a look around. You should see flour spread out for a long way on your floor, blown outward by your breath.

However, you'll see that some flour was carried farther away than your breath alone could have blown it. It's hard to get a good breeze blowing as far away as your outstretched hand because your breath can really only go a few dozen centimeters before petering out. What carries the dust farther than your breath can go is the air that already exists in the room. You blew air from your lungs, and that air displaced the air in the room, and it was *that* air that carried the flour farther than your breath alone could push it.

However, on the Moon, *there is no air*. The thrust of the LM engine was substantial, but it only blew the dust out from directly beneath it. Some of that dust blew for hundreds of meters, but, contrary to our experience here on Earth, the dust just *outside* the immediate area where the exhaust plume touched down was largely left alone. Plenty of dust was left there in which to leave footprints. In reality, a little more dust got blown around than that because the dust blown around directly by the engine can smack into other particles of dust, moving them also. So the "hole" in the dust was bigger than the burn area of the rocket, but not substantially so. Incidentally, in the tapes of the Apollo 11 landing you can hear Buzz Aldrin commenting that they were "picking up some dust" from the engines as they neared the surface. Neil Armstrong, who piloted the LM, complained that the moving dust made it hard for him to figure out how fast they were moving across the surface.

Some hoax-believers also claim that the dust could not keep footprints because it has no water in it, and you need something wet to keep it compacted. This is nonsense. Baking soda, for example, is incredibly dry, yet you can easily leave a footprint in it. This claim is bizarre, and again I am dumbfounded as to why someone would put it forward when it is so trivially easy to prove wrong by experiment. In this case, at least, common sense leads you the right way.

4. The Temperature of the Lunar Surface

Related to the dust problem is that of the Moon's temperature. The Apollo missions were made during the day on the Moon. Measurements of the Moon's surface show that the temperature can get as high as 120°C, hot enough to boil water! Hoax-proponents point out that the astronauts could not have lived through such fierce heat.

In one sense they are correct: that much heat *would* have killed the astronauts. However, the astronauts were never in that much heat.

The Moon spins on its axis once every 27 days or so. That means that a lunar day is four weeks long, with two weeks of sunlight and two weeks of darkness. Without an atmosphere to distribute the heat from the incoming sunlight, the daylit side of the Moon does get tremendously hot, and the dark side gets very cold, as cold as $-120°C$.

However, the surface doesn't heat up the instant the sunlight touches it. At sunrise the sunlight hits the Moon at a very low angle, and it does not efficiently heat it. It takes days for the lunar surface to get to its high temperature, much as the worst heat of the day on Earth is reached after the Sun reaches its peak. NASA engineers, knowing this, planned the missions to take place at local morning, so that the Sun was low in the sky where they landed. You can see this in every photograph taken from the surface; the shadows are long, indicating the Sun was low in the sky.

As it happens, the spacesuits *were* designed to keep the astronauts cool, but not because of the outside heat. In a vacuum, it's very difficult to get rid of the astronauts' own body heat. An astronaut inside an insulated suit generates a lot of heat, and that heat needs to get dumped somehow. The suits needed ingenious methods to cool the astronauts. One way was to pipe cool water through tubes sewn into their undergarments. The water would warm up, picking up their waste heat, then flow into the backpacks where the heat could be dumped away into space.

So there really was a problem with temperature, but it was internal, not external. Another hoax claim, frozen in its tracks.

Incidentally, the dust on the surface of the Moon is a terrible conductor of heat. Powdery materials usually are. Although the dust was actually warmed by sunlight, it wasn't able to transfer that heat well to the astronauts through their boots. Oddly, even though the surface of the Moon gets to 120°C at noon, the dust is only that hot for a short distance down, because the heat can't flow well below that depth. Below that depth, the rock is eternally freezing cold, insulated by the dust and rock above it. The dust cools rapidly once the Sun sets. During a lunar eclipse, when the Moon is in the Earth's shadow, the lunar temperature has been measured to drop very quickly. The dust gets as cold as the rock beneath it.

That coldness came back to haunt one astronaut. During an excursion, Apollo 16 astronaut John Young realized that the rocks they had collected were all rather small. He wanted one really big one to impress the scientists back home. He grabbed a rock weighing roughly a kilogram (two pounds), and placed it underneath the lander, in shadow, while he closed up shop to prepare for the return to Earth. When he was done, he put the rock on the LM and repressurized the module.

It was then that Young realized he needed to rearrange the rocks a bit to balance them in the LM, making sure that the spacecraft wouldn't tilt dangerously during takeoff due to an imbalance in the mass distribution that the automatic controls couldn't handle. He had already taken his gloves off, and when he grabbed the big rock, he got a surprise: the rock had been in the shadow long enough to dump its extra heat, and had become bitterly cold! Young was actually lucky not to get frostbite. When Young retold this story to Paul Lowman, a NASA geologist and lunar expert, Lowman exclaimed, "This is the only time I've ever heard somebody describe the actual temperature of the Moon as he actually felt it!"

Hoax-proponents also claim that the film carried by the astronauts would have melted in the tremendous lunar heat. In reality, the opposite problem is true: they didn't have to worry about film melting; they had to insulate it to keep it from freezing.

5. *Tricks of Light and Shadow*
Another common line of "proof" of a NASA conspiracy has to do with the play of light and shadows on the Moon. The most common of these claims concerns the blackness of the shadows. If the Sun is the only source of light, say the hoax-believers, then shadows should be absolutely black because there is no scattered

sunlight from the air to fill them in. Without any light illuminating the ground in the shadow, it should be completely, utterly black.

On the Earth, we are accustomed to shadows that are not actually totally black. This is due primarily to our bright sky. The Sun itself casts a sharp shadow, but the light from the air in the sky illuminates the ground in our shadow, making us able to see objects there.

On the Moon, where the sky is black, conspiracy theorists claim the lunar surface inside the shadow should be completely black. If the Sun is the only source of light, they say, the shadows should be black as pitch. Yet, in the astronaut photographs we commonly see shadows filled in a bit, as if there were another source of light. Obviously, to the hoax-proponents, since the Apollo photographs were taken on a soundstage on Earth, the source of this light is the air inside the building, scattering the light from a spotlight.

However (stop me if you've heard this before), they're wrong. There *is* a source of light on the Moon besides the Sun, and we've already said what it is: *the Moon*. The sky may be black, but the surface of the Moon is very bright and reflects the sunlight, filling in the shadows. This is another trivially simple answer to one of the hoax-proponents' "puzzling" questions.

Interestingly, sometimes the shadows falling on the lunar surface appear to be filled in as well. Ironically, the source of light is most likely the astronauts themselves. The spacesuits and the LM are brightly lit by the Sun and the lunar surface, and that light is reflected back onto the lunar surface, filling in the shadows a bit. This exact same technique is used by photographers and cameramen, who employ umbrella-like reflectors to fill in the shadows when photographing a scene.

However, if you look more closely at the photographs, the problem does get more complicated. In what has become the most famous picture taken on the Moon, Neil Armstrong snapped an image of Buzz Aldrin standing near the LM during the Apollo 11 mission (see page 169). We see Buzz facing the camera, lit by the Sun from behind and to the right. Reflected in his helmet we can see Neil's image as well as the lander leg and various shadows.

One of the most famous photographs from the Apollo missions, the "Man on the Moon" picture of Buzz Aldrin. Conspiracy theorists point to many clues that indicate the image was faked: the lack of stars, the filled-in shadows, and the apparent spotlight effect. However, all of this is in fact evidence that the picture is genuine. Note also Aldrin's knees; they are covered with ash-gray lunar surface powder from the many times Buzz had to dip down to pick up a dropped tool or collect a rock sample. Despite what others might say, this image was indeed taken on the surface of an alien body, the Earth's Moon. (Photograph courtesy of NASA.)

This image is of paramount importance to the hoax-believers. It embodies two claims critical to their arguments: From the way the ground is illuminated Aldrin is clearly being lit by a spotlight aimed directly at him, and from shadows in his visor it looks as if that spotlight is nearby.

This picture is oddly lighted, but not because of any human trickery. Actually, the spot of light results from a peculiar property of the lunar surface: it tends to reflect light back in the direction from which it comes. This is called backscatter, and it is very strong on the Moon. If you were to shine a flashlight in front of you there, you'd see the light strongly reflected back to you. However, someone standing off to the side would see hardly any reflected light at all.

Actually, you've almost certainly seen this effect on your own. You might guess that the half-full Moon is half as bright as the full Moon, but that's not correct. The full Moon is roughly *ten times* as bright (H. N. Russell, "On the albedo of planets and their satellites,"

Astrophysical Journal 43 [1916]: 103). That's because at full Moon, the Sun is shining from directly behind you, straight onto the Moon. The lunar soil then obligingly sends that reflected light straight back to you. At half Moon, the light is coming from the side and much less is reflected in your direction, making the Moon look fainter.

That's why Aldrin appears to be in a spotlight. In the area where he's standing, the light is reflected straight toward Armstrong's camera. Farther away from Aldrin, though, the light gets reflected *away* from the camera, making it look darker. The effect generates a halo of light around Aldrin.

The technical name for this glow is **heiligenschein,** which is German for "halo." You can see it yourself on a dewy morning. Face away from the Sun so that the shadow of your head falls on some wet grass. You can see the glow of backscattered sunlight surrounding your shadow's head, looking very much like a halo. You can also do this where the ground is dusty, such as in a baseball diamond infield. The effect can be very striking. This "spotlight" effect can be seen in many Apollo photographs, but only when the astronaut taking the picture had his back to the Sun, just as you'd expect. There's no spotlight, just some odd—but natural—physics at work.

Incidentally, the opposite effect happens when you drive a car on a rainy night. Wet pavement reflects the light *forward,* away from you. Oncoming cars can see your headlights reflected in the pavement, while your headlights hardly seem to light up the road in front of you. The light is thrown ahead of you, not back at you, making it hard to see the road.

The second claim about the photograph deals with the shadows. If you look in Aldrin's visor, you'll see that the shadows aren't parallel. If the Sun is the source of light, all the shadows should be parallel. Instead, they point in different directions, which means the source of light must be close by. Ergo: it's a spotlight.

Well, we've already seen it's not a spotlight, so we know it must be the Sun. Actually, this claim is another ridiculously easy one to refute. We see the shadows reflected in a *curved* visor. The curvature of the visor distorts objects in it, like a fisheye lens or a funhouse mirror. The shadows are curved because the visor is

The shadows cast by the astronauts, rocks, and other surface features appear to be nonparallel, but this is just an effect of perspective, similar to the apparent converging of railroad tracks on the horizon.

curved. That's all there is to it. Again, no fakery, just simple optics that everyone has seen at some point in his or her life.

However, there are also some images that are not visor reflections, but still seem to have shadows pointing in different directions. Again, if the Sun is the only source of light, shadows should be linear and parallel. Clearly, sometimes they are not parallel. To the conspiracy theorists, of course, this is more evidence that the images are fake.

Have you ever stood on a set of railroad tracks and seen how they appear to converge far away, near the horizon? This is an effect of perspective, of course. The railroad tracks are parallel (they wouldn't be much use if they weren't), but our eyes and brain interpret them as converging.

The same thing is happening in the lunar photographs. The shadows don't appear to be parallel because of perspective. When comparing the directions of shadows from two objects at very different distances, perspective effects can be quite large. I have seen this myself, by standing near a tall street lamp around sunset and comparing its shadow to that of one across the street. The two shadows appear to point in two very different directions. It's actually a pretty weird thing to see.

Again, this is something that can be investigated quite literally in your front yard, and is hardly evidence of a multibillion-dollar conspiracy.

🐌🐌🐌

There's an interesting lesson here about the claims of the hoax-believers.

In many cases they use simple physics and common sense to make their points. Usually their initial points make sense. However, they tend to misunderstand physics, and common sense may not apply on the airless surface of an alien world. Upon closer inspection, their arguments invariably fall apart.

I could go on and on with more examples. Debunking the hoax-believers' claims could fill a book. That's not surprising, considering several books have been written by them. I have no doubt the books sell well. Conspiracy books always do. I also have no doubt that a book dedicated to debunking them would *not* sell well. A whole book pointing out the believers' errors would be tedious, and it isn't necessary. The examples above are the strongest they can muster, and they fall apart easily when shaken. Their other arguments are even weaker.

But the interesting part is the seeming simplicity of their claims. Not seeing stars in the Apollo pictures is so obvious, so basic a mistake. The other arguments they make seem obvious as well.

But let's a have small sanity check here. Let's say NASA knew it couldn't put men on the Moon, and knew it would lose all its money if it didn't. They decided to fake the whole lunar project. They built elaborate sets, hired hundreds of technicians, cameramen, scientists knowledgeable enough to fake all this, and eventually spent millions or billions of dollars on the hoax. Eventually, they put together the greatest hoax in all of history, *yet they forgot to put stars in the pictures?*

There's more. It has come to light in recent years that the Soviets were well on their way to sending men to the Moon in the 1960s as well. Their missions never got off the ground, but the Soviets worked very hard on them, and of course they were watching carefully when NASA broadcast its own footage. Both superpowers had spent billions of dollars on their respective lunar projects; national prestige was at stake for the two countries that just a few years before were on the verge of nuclear war. You can imagine that if the Soviets had faked their missions and forgotten such obvious flaws as stars in images and shadows that went in the wrong direction, the American press would have savaged them beyond belief. Do the conspiracy theorists honestly think that *Tass* or *Pravda* would have done any differently to the American project? It would have been the Soviets' greatest victory of all time to prove that the Americans had botched their biggest peacetime project in history, yet even *they* acknowledged the truth of the Moon missions.

In the end, truth and logic prevail. America did send men to the Moon, and it was triumph of human engineering, perseverance, and spirit.

❧ ❧ ❧

A postscript: after Kaysing finished his book *We Never Went to the Moon,* he approached Jim Lovell with it. Lovell was the commander of Apollo 13, and literally came close to death trying to save his crew and his ship after an explosion crippled the spacecraft. Lovell's stake in the space program is almost beyond comprehension.

So you can imagine Lovell's reaction when he read Kaysing's book. In the *San José Metro Weekly* magazine (July 25–31, 1996), he is quoted as saying, "The guy [Kaysing] is wacky. His position makes me feel angry. We spent a lot of time getting ready to go to the moon. We spent a lot of money, we took great risks, and it's something everybody in this country should be proud of."

Kaysing's reaction to Lovell's comments? He sued Lovell for libel. In 1997, a judge wisely threw the case out of court. There's still hope.

18

Worlds in Derision: Velikovsky vs. Modern Science

In 1950, a remarkable book entitled *Worlds in Collision* was published. It was the culmination of a decade's work by a man who had a startling thought: what if the various disasters recorded in ancient texts were real, actual events?

The ancients experienced so many catastrophes that it almost sounds like something from a bad science fiction movie. Fire rained down from the sky, the Sun stood still during the day, floods, famines, vermin infestations—it seems like things were a bit more exciting back then. Of course, most people assume that these events were either exaggerated or were simply myths spawned from storytelling and a very human need to explain things that are beyond our understanding. But suppose we take these ancient writers at their word, and assume that these events really did happen. Can there be a simple, common cause? Could it have an astronomical basis?

Psychoanalyst Immanuel Velikovsky decided to tackle this issue. His answers to these seemingly simple questions would have massive repercussions throughout the scientific community, although perhaps not in the way he would have thought. By the time he finished *Worlds in Collision* and its sequel, *Earth in Upheaval,* he honestly felt that he had uncovered evidence that all previous scientific laws were wrong, and that we needed to seriously rethink the way the universe worked.

Many people avidly read *Worlds in Collision,* putting it on the bestseller list soon after it came out. It was a counterculture smash in the 1960s and '70s. His popularity is waning now, but Velikovsky still has many followers, many of whom fiercely defend his notions.

What are these notions? The basic premise espoused by Velikovsky is that the planet Venus was not formed at the same time as the other planets in our solar system. Instead, he concludes it was formed recently, only a few thousand years ago, around the year 1500 B.C. According to Velikovsky's analysis of the Bible and other ancient tomes, Venus was originally part of the planet Jupiter, which somehow split apart, ejecting Venus bodily as a huge comet. Over the ensuing several centuries, Venus careened around the solar system, encountering the Earth and Mars multiple times, affecting them profoundly. It was the gravity and electromagnetic effects of these near passes of Venus and Mars as well that caused all the catastrophes heaped upon our ancestors.

As you might guess, I disagree with Velikovsky. I'm not alone; nearly every accredited scientist on the planet disagrees with him as well. There's good reason for this: Velikovsky was wrong. Really, really wrong. The astronomical events he describes are not so much impossible as they are *fantastically* impossible—literally, they are fantasy.

To be fair, a lot of accepted scientific theories sound fantastic, too. Who can believe that the universe started as a tiny pinpoint that exploded, creating time and space, which then began to expand, forming the cosmos as we see it now?

What you have to remember is that the Big Bang was first proposed after many astronomical observations were made that could not be explained any other way. There has been a great deal of support for the Big Bang for decades now, and it's actually one of the most solid ideas in science. On the other hand, Velikovsky's ideas have little support from astronomical observations, and in fact many fairly well-established astronomical theories directly contradict his ideas. The difference between the Big Bang and Velikovsky's thesis is physical evidence. The former has lots, the latter has none.

Velikovsky's thesis certainly *seems* legitimate. It's built upon a tremendous amount of historical and archaeological research. The book has a vast number of quotes from all manner of historians, from contemporary analyses to those by the ancient Roman, Pliny the Elder. Experts in the field have many criticisms of Velikovsky's interpretations of these works, and it's quite possible that his research is historically inaccurate. To be honest, I have no expertise in this, and so I'll refrain from judging his ideas on their historical merit. However, I'll be happy to discuss them in an astronomical context.

Like most areas of pseudoscience such as astrology and creationism, it's possible to find fatal flaws in the theories without resorting to a detailed and painful analysis of every fact and figure. As a matter of fact, sometimes it pays not to nitpick; when you do, pseudoscience supporters will simply throw more facts and figures at you, hoping either to dazzle you with their database of knowledge or to confuse you beyond hope of reaching any rational conclusion. So, instead of going over his writings with a fine-toothed comb, it's a better idea to look at more general concepts—large, broad areas that contradict the basic premise. These are usually easier to explain and understand, anyway. If I mention a *few* details it's because I think they're important, as well as fun and interesting.

❧ ❧ ❧

Velikovsky's main idea of Venus as tooling around the solar system and creating havoc is based on many ancient writings. Perhaps the most important biblical passage is in Joshua 10:12–13. During a massive battle with the Canaanites, Joshua knew he could win if only he had a little more time, but the day was drawing to a close. Getting desperate, he asked God to make the Sun stop its daily motion around the Earth, giving him the extra time he needed. The biblical passage reads, ". . . and the Sun stood still in the midst of heaven, and the Moon stayed, until the people had avenged themselves upon their enemy." Then, almost exactly 24 hours later— after the battle was over—God restarted the heavens, setting the Sun and Moon in motion once more.

Today, we would interpret that—if we are particularly literal-minded—to mean that the Earth's rotation stopped, so that it only appeared that the Sun and Moon were motionless in the sky. Somehow, a day later, it started its rotation again.

Velikovsky, quite literal-minded indeed, was researching this event and discovered reports of a meteor storm that supposedly happened just before the Earth's motion stopped. To him, the meteors indicated an astronomical cause for the biblical passage. This idea dovetailed neatly with other legends he found, such as those in ancient Greece. The goddess Minerva, associated at the time with the planet Venus, was born fully grown from the head of Zeus (associated with the planet Jupiter). Other cultures had vaguely similar claims of ties between Jupiter and Venus. Velikovsky suspected that these legends were actually based in fact. From there, he shaped his idea that the planet Venus was indeed literally ejected from the planet Jupiter, and subsequently encountered the Earth on multiple occasions.

It was the first such encounter with Venus that stopped the Earth's rotation. Somehow—Velikovsky is never really clear on this, but instead invokes vague claims of a previously unknown electromagnetic process—Venus was able to slow and stop the Earth's spin during an exceptionally close pass. Venus then moved off, but a day later came back for a second pass that started the Earth's rotation again. Venus itself was sent off in a long, elliptical orbit, only to pass by the Earth again some 52 years later. Over time, Venus settled down into its present orbit as the second planet from the Sun.

There are so many flaws with this idea that it's difficult to know where to start. For example, Velikovsky points to many passages in ancient texts that describe a great comet in the sky, the passing of which precedes many of these catastrophes. How to reconcile this with the planet Venus? Well, he says, Venus was ejected from Jupiter *as a comet*. It didn't become a planet proper until it found its way into a stable orbit around the Sun.

First, ejecting something with the mass of Venus would be very difficult to say the least. Velikovsky suggests that Venus fissioned off, flung outwards by Jupiter's rapid rotation in the same way that a dog shakes its body to spray off water after a bath. In reality, this won't work.

There are many lines of evidence showing the solar system to be billions of years old. Why would Jupiter wait until a few thousand years ago—a tiny, tiny fraction of its lifetime—to suddenly eject a planet-sized mass? The only way to shrug off this huge coincidence is to say that this is not a rare event, and that Jupiter has done it many times before. But where are these planets? If you assume all the planets formed this way, then you are left with the problem of how Jupiter formed. Since Jupiter would slow its own rotation every time this happened, it would have had to start with an impossibly high rotation rate.

Second, Venus and Jupiter have entirely different compositions. Jupiter is mostly hydrogen, the lightest element. It probably has denser elements in its core, but Venus should show at least *some* similarities to Jupiter. However, they're about as different as two planets can be. The chemical composition of Venus is very much like that of the Earth, so it seems unlikely that similar planets would have formed in such vastly different ways.

Third, Venus is a fair-sized planet. In fact, it has almost exactly the same mass and diameter as the Earth. Jupiter is orbited by a retinue of moons, four of which are so big that they would be planets in their own right if they didn't orbit their mighty host. These moons orbit Jupiter in nice, almost perfectly circular orbits, which is what one would expect after millions or billions of years of gravitational interaction with Jupiter and each other. (See chapter 7, "The Gravity of the Situation," for more about tidal evolution.)

Now imagine Venus plowing outward through this system. The orneriest bull in the most delicate of china shops would be nothing compared to the devastation wrought on the jovian system. The moons would get scattered, their orderly orbits perturbed by the rampaging planet on its way out bent into long ellipses. Some may even have been ejected from Jupiter completely to wander space as Venus reportedly did; yet there is no mention of these rogue moons in ancient texts.

We see no evidence of this at all in the system of moons around Jupiter. By all observations, the moons have been doing what they do now for the past billion or two years at least. If there have been any disturbances, they *certainly* have not occurred in the past few millennia.

Velikovsky spends quite some time in *Worlds in Collision* trying to show that Venus was ejected by Jupiter. He is wrong; such an event simply could not happen. It can be shown mathematically that the amount of energy needed to eject Venus would have literally vaporized the planet! In other words, whatever type of event Velikovsky envisioned to shoot Venus out of Jupiter actually would have turned Venus into a very hot, incandescent gas, exploding outwards like, well, an *explosion*. It certainly would not have formed a solid body able to roam the solar system. This seriously weakens his argument about Venus wandering the solar system, unless you believe in truly immense forces having incredibly benign effects. It would be like dropping an anvil on an egg and winding up with two perfectly split egg shells, one with the white inside and the other with the yolk. When forces are that huge, they rarely clean up after themselves so neatly. In reality, the egg would be a goopy mess, just as Venus would be by whatever forces Velikovsky was imagining.

Still, let's grant that some mysterious unknown force set Venus in motion. So, ignoring the genesis of Venus, is it still possible that it somehow passed so close to the Earth that it caused widespread disaster here?

In a word: no.

From his readings Velikovsky concludes that Venus passed close enough to the Earth to stop its rotation, moved off, then came back a few hours later and started the Earth moving again. However, he is very vague about the exact mechanism for this. He theorizes that perhaps the Earth didn't really slow and stop, but instead it flipped over on its axis, making the north pole become the south pole and vice versa.

Indeed, he spends dozens of pages giving evidence that the Earth didn't always spin with the north pole in the position it is now. He starts a chapter about this as follows: "Our planet rotates from west to east. Has it always done so?" He quotes ancient texts as saying that the Earth has flipped not just once, but many times.

His basis for this is pretty shaky. One passage he quotes talks about two drawings of constellations found in an Egyptian tomb.

In one drawing, the constellations are represented correctly, and in the other they are reversed east-to-west, as if the Earth were spinning the wrong way. How else to account for this but to assume the Earth indeed was spinning east to west?

Actually, there are two ways. One is that the Earth is a big ball. To someone standing in the southern hemisphere, the constellations will look upside down compared to the view of someone standing in the northern hemisphere. The curvature of the Earth does this, making one person look like he is "standing on his head" relative to another. That would explain the upside-down constellations pretty well; perhaps a traveler was describing what he saw Down Under.

There is another explanation as well. Many ancients thought stars were holes in a great crystal sphere, letting the light of heaven shine through. The gods lived on the other side and therefore saw the constellations backward relative to us. Many star maps show this so-called "gods' view" of the sky. In the main corridor of Grand Central Terminal in New York City, the stars are painted on the ceiling this way. Perhaps that Egyptian drawing was showing our view of the sky versus the gods' view.

I find either of these explanations a bit more palatable than calling for the Earth to flip over.

And again, even if we do grant that the Earth flipped over, Velikovsky would have us believe that *another* pass of Venus 24 hours later flipped the Earth back the way it was, and spinning at the same rate. To put it *very* mildly, this is pretty unlikely.

❧ ❧ ❧

There are still two more massive, basic, truly fatal flaws to Velikovsky's Venus theory: one is that we still exist, and the other is that the Moon is still around.

Velikovsky goes to great pains, over hundreds of pages in his book, to relate the various disasters that befell mankind as Venus loomed hugely in the sky. All of these events call for Venus to get pretty close to the Earth. At one point, to explain such things as manna falling from heaven and the Egyptian plague of vermin, he states that Venus gets so close that its atmosphere flows into the Earth's own air.

This premise, that manna and insects came from Venus to the Earth, is suspect at best. We now know that the surface of Venus has an incredibly high temperature, over 470° Celsius (900° Fahrenheit), hot enough to melt lead. It's difficult to imagine what kind of bug could survive such withering heat. It's also hard to see how manna—a life-sustaining compound—could form on Venus. After all, the atmosphere of Venus is mostly carbon dioxide and sulfuric acid. If a few billion tons of these substances gets dumped into the air here on Earth, the effect would hardly be conducive to life. Quite the opposite.

There are other physical effects of a Venusian near miss. Despite its differences, Venus does have some similarities to the Earth. They have almost exactly the same mass and diameter. That means they have about the same gravity. For the air of Venus to flow onto the Earth, there would need to be about an equal pull from both planets on the air, with a little bit more of a pull from the Earth. Even being outrageously generous, their nearly equal gravity means that Venus would have to be closer than 1,000 kilometers (600 miles) from the surface of the Earth.

Imagine! A planet the size of the Earth passing just 1,000 kilometers overhead would be just about the most terrifying event I can imagine. Venus would literally fill the sky, blocking out the Sun and stars. Even at interplanetary velocities it would be vastly huge in the sky for days or weeks, and would be brighter than hundreds of full Moons.

Yet, no mention of this incredible spectacle is made in ancient texts.

Worse, the tides from Venus on the Earth would be huge, kilometers high. The earthquakes would have been more than terrible; they would have destroyed everything, and I mean *everything*. It would make the sweatiest vision of biblical apocalypse look like a warm spring day. That close a pass by something like Venus would have sterilized the surface of the Earth, killing every living thing on it. Had it happened, Velikovsky wouldn't have been around to write his book. To his credit, Velikovsky assumes that there would have been earthquakes and the like, but he underestimates the effect of such a near passage by a factor of millions.

Even granting that humanity somehow survived this apocalypse, there is *still* a problem. The Moon orbits the Earth at a distance of 400,000 kilometers. If Venus were to get so close to the
Earth that it could actually exchange atmospheric contents, it
would have to get closer to us than the Moon. If that had happened, the Moon's orbit would be drastically changed. Usually
when you take three objects, two of them massive and one less so,
and let them interact gravitationally, the least massive one is
ejected from the system completely. In other words, under almost
every circumstance, had Venus come that close to the Earth, the
Moon would have literally been flung into interplanetary space. At
the very least its orbit would have been profoundly changed, made
tremendously elliptical.

The orbit of the Moon *is* elliptical, but nowhere near as stretched
out as it would have been after a close encounter with Venus. The
very fact that the Moon—and we, too—are here at all shows that
Velikovsky was wrong.

There's still one more thing. The Hebrew calendar, invented
2,400 years ago, is based on lunar cycles. The events described by
Velikovsky supposedly took place only a thousand years before the
Hebrews invented their calendar. But any passage by Venus would
have radically changed the Moon's orbit, and 1,000 years isn't
nearly enough time for the Moon's orbit to settle down into what
it is now. In fact, the month has not changed length appreciably
since the Hebrew calendar was invented, meaning the Moon would
have had to settle down almost instantly after Venus swept by.

Remember, Velikovsky used ancient texts to support his beliefs.
Yet, here we see that one of the most basic of all ancient tools, the
calendar, *refutes* his hypothesis. Had Venus done any of the things
Velikovsky claimed, the Moon's orbit would have changed, and we
see no change.

❦ ❦ ❦

Finally, according to Velikovsky, after Venus was done palpating
and poking the Earth, it finally calmed down and settled into its
present orbit. Remember, if this happened it's far more likely that

Venus' orbit would be a highly eccentric ellipse rather than a circle. Yet Venus has the most circular orbit of all the planets, matched only by Neptune. If we accept that Venus actually went through all these Velikovskian gyrations, we would at least expect it to have some mild eccentricity, yet Venus' orbit is difficult to distinguish from a perfect circle.

Later in his book, Velikovsky tried to explain how all the orbits of the moons and planets could have become more circular. He proposed that there is an electromagnetic force that emanates from the planets and the Sun, and indeed it is this force that flipped the Earth over and created all the problems. However, today we see absolutely no evidence *whatsoever* that such a force exists in anywhere near the strength needed to do a *fraction* of what he claims. If such a force ever existed, it stopped working shortly after the events described in the Book of Exodus. Also, if such a force existed, why are some planets' and moons' orbits *not* perfect circles, or even close to them? We see some objects on highly elliptical orbits; comets are a good example. Why didn't this force affect them?

So, Velikovsky would have us believe that all these biblical disasters occurred due to some mysterious force, unknown in nature, that came to the planets, did its nefarious deeds—but only on some objects, ignoring others—and then evaporated again. It left not a trace on the planets, either, as the solar system looks exactly like it had evolved naturally over billions of years.

This is not science. In fact, it would be just as legitimate to invoke the Hand of God. In other words, there is hardly a need for Velikovsky to go to such great lengths to try to get science to corroborate his beliefs. His use of a force unknown to science negates the entire purpose of writing the book in the first place, which was to find a scientific grounding for ancient texts.

❧ ❧ ❧

So if Velikovsky was so utterly and obviously wrong, why do so many people still follow his work and think he was right?

This question is more philosophical than practical. However, part of the answer lies in the way the scientific community treated Velikovsky when he published his book.

Initially, in 1950, when the Macmillan publishing house was preparing the manuscript for publication, the scientific community caught a whiff of it. In particular, a Harvard astronomer named Harlow Shapley wrote several vitriolic letters to the editors at Macmillan saying—correctly, mind you—that Velikovsky's ideas were wrong, and that Macmillan was doing everyone a big disservice by publishing them. At the time Macmillan was a very large publisher of scientific textbooks, and Shapley said that the publisher's reputation would be damaged by selling *Worlds in Collision*. From what I have read, there were intimations, although not direct threats, that Shapley would use his considerable reputation to pressure other scientists to boycott Macmillan's books.

This was a serious problem for the publisher. When Velikovsky's book came out it rocketed onto the bestseller list, no doubt aided by the controversy. It was a huge money maker. Macmillan, however, also made a lot of money from textbooks. In one of the worst publishing decisions ever, bowing to the pressure, they transferred the rights to *Worlds in Collision* and its sequels to Doubleday, which suddenly found themselves printing a book they couldn't keep on the shelves. This only added to the book's mystique, aiding its sales.

With sales booming, the scientific pressure against Velikovsky continued. His book became a favorite among college students, especially in the 1960s when intellectual rebellion was fashionable. The situation became so bad, as far as "establishment" scientists went, that the American Association for the Advancement of Science sponsored a semipublic debate in 1974 between Velikovsky and his detractors in an attempt to discredit the book once and for all. One of the leading scientists in the debate was Carl Sagan, who by then was something of a media darling, a professional skeptic, and well-known by the general public.

I did not attend this debate, as I was only nine years old at the time. However, I have read many accounts of this infamous meeting on the web and in books. Which side won, Velikovsky's rebels or mainstream science? In my opinion, neither. I'd say they both lost. Velikovsky made several rambling speeches that neither supported nor detracted from his cause, and his supporters came across more as religious zealots than anything else. On the side of science, there was much posturing and posing. Sagan—for whom I have tremendous respect both as a scientist and as someone who popularized teaching astronomy to the public—did a terrible job debunking Velikovsky's ideas. He made straw-man arguments, and attacked only small details of Velikovsky's claims.

The book *Scientists Confront Velikovsky* [Cornell University Press, 1977] transcribes the talks given by scientists at the meeting. As it happens, Velikovsky's talk is not in the book. Sagan was given an extra 50 percent more space to rebut Velikovsky's arguments using arguments not in Sagan's original paper, but Velikovsky was not given any room to counter Sagan's rebuttals. Because of this word-length dispute, Velikovsky withdrew his paper from the book. In the book, Sagan gives his arguments against Velikovsky and expands upon them even further in his own (otherwise excellent) book, *Broca's Brain*. Again, Sagan's arguments are not all that great. For example, he gives the energy criteria necessary for Jupiter to eject Venus but then ignores Jupiter's own rotation, which is crucial for the analysis. On his web site about the affair, fellow scientist and author Jerry Pournelle calls Sagan's performance "shameful."

Sagan's and Shapley's reactions were not uncommon in any way among scientists. Many of them loathed the very idea of Velikovsky writing this book and the fact that he was getting rich from it too. But the extreme amount of bile and bitterness only helped make Velikovsky a martyr. To this day he is practically revered by his followers.

Thomas Jefferson once wrote, "I know no safe depositary [sic] of the ultimate powers of the society but the people themselves; and if we think them not enlightened enough to exercise their control with a wholesome discretion, the remedy is not to take it from them, but to inform their discretion by education." Perhaps, if Shapley and his fellow scientists had heeded Jefferson, *Worlds in Collision* would be just another silly pseudoscientific book collecting

dust next to ones about UFO aliens curing pimples using homeo-
pathic crystals. Instead, even after half a century, it can be found
on bookshelves today.

❧ ❧ ❧

There's an ironic footnote to this episode in the history of science.
Certainly, scientists of the day dismissed Velikovsky because his
assertions clearly flew in the face of everything known about physics
and astronomy, then and still today. They also ridiculed him because, at the time, it was thought that the planets were fairly static.
Things didn't change much. Any change that occurred was gradual, slow, glacial. Nothing happened suddenly. This type of thinking is called **uniformitarianism.**

However, this tide was turning. As observations of the planets
improved, including our own, we started to learn that things didn't
always happen at a stately rate. The Moon is covered with craters;
it was once thought that these were volcanic, but around the same
time as *Worlds in Collision* was published, scientists were starting
to speculate that at least some lunar craters were formed from
meteor impacts. Venus' surface bears evidence that some massive
event resurfaced the whole planet some hundreds of millions of
years ago, and it looks like there have been many mass extinctions
caused by individual catastrophic events here on Earth.

Today we understand that both uniformitarianism and catastrophism describe the history of our solar system. Things mostly
go along slowly, then are suddenly punctuated by rapid events.

Velikovsky supporters claim that he was simply ahead of his
time, and his theories of catastrophism were denied their due. This
is silly; just because he used the idea that catastrophes happened
doesn't mean that any of the things he described were right. But it
is rather funny that scientists of the day were wrong in many of
their assertions of uniformitarianism as well.

Still, that's the difference between science and pseudoscience:
scientists learn from their mistakes and abandon theories that don't
pan out. Velikovsky was wrong, as were the scientists at the time.
But science—real science—has moved on. Maybe we can all learn
something from this.

19

In the Beginning:
Creationism and Astronomy

There is a story, almost certainly apocryphal, about a scientist who was giving a public lecture on astronomy. He was describing the scale of the universe, starting with the Earth orbiting the Sun and working his way up to galaxies orbiting other galaxies, and eventually the structure of the universe as a whole. When he was done, an old lady stood up and challenged him.

"Everything you just said is wrong," she claimed. "The Earth is flat, and sits on the back of a giant turtle."

The astronomer knew immediately how to retort to *that* statement: "But then, dear lady, on what is that turtle standing?"

She didn't bat an eye. "You're clever, sir, very clever," she said. "But it's turtles all the way down!"

I've always liked that story. Most people think it's about a silly old woman who doesn't understand anything at all about science. But I wonder. It's not too hard to play a little role reversal. After all, is her answer any more silly than saying that the universe started out as quantum fluctuation that caused the violent expansion of space-time itself?

Okay, yes, it is sillier. But the scientific explanation of the universe, although steeped in observation and tempered through the scientific method, may sound pretty silly to someone not well versed in the subject. The philosopher Pierre Charron said, "The true science and the true study of man is man." But in a very real sense,

man is a part of the universe. I think, after 400 years, we can up-
date Charron's statement: The true science and the true study of
man is the universe.

We've been asking basic questions about our existence for a
long time. Why are we here? Does the universe have meaning?
What is our place in it? How did it all begin? These are questions
of the most fundamental nature that everyone asks at some time.
People turn to all manner of oracles for answers—religion, science,
friends, recreational drugs, even television, although TV usually
raises more questions that it answers.

That last question is the real poser. How *did* the universe
begin? Everything in our lives has a beginning and an end. Stories
start, build, reach a climax, then finish. Pictures have borders, sym-
phonies have first and last movements, vacations have a starting
and stopping point. Of course, our lives themselves are framed by
birth and death. We experience everything one second at a time, an
orderly flow from early to middle to late. We expect the universe
to reflect our conditioning, that it had a beginning and that it, too,
will eventually end.

Of all the philosophical questions, this one may actually have
some scientific meaning. The clues to the beginning are there, if we
can decode them. The universe is like a giant book, and if we are
smart enough, we can turn the pages and read it.

To push this analogy just a little further, the next question
might be, "In what language is the universe written?" This question is at the heart of a lot of debate. It may not surprise you that
I think the universe follows a set of rules—physical and natural
rules. These rules are complex, they are not clear, and it is beyond
doubt that we do not understand all of them, or have even imagined what they could be. Some are simpler, like the behavior of
gravity. Others are complex beyond our mind's capability, like how
matter can disappear down a black hole, or just why an electron
has a negative charge. But no matter how simple or how complex,
the language of the universe is physics and math. We learn this language better as we observe phenomena around us.

Some people, though, do not think this way. They *presuppose*
a set of rules and try to get their observations to match what they

want to believe. This isn't a great way to try to figure out the universe. You wind up having to throw out observations that don't match your beliefs, even if those observations are showing the universe's true nature.

Such is the case for a minority sect of Christianity who call themselves Young Earth Creationists. This is a vocal minority, however, and their cries are heard loudly in the United States. They believe that the Christian Bible is the inerrant Word of God, accurate in every detail and the only way to judge observations. They believe that anything that does not agree with what is written in the Bible is wrong. Moreover, the Bible is not really up for interpretation: what it says, goes.

One faction of the young-Earth crowd is the vociferous Institute for Creation Research, or ICR. They are dedicated "to see science return to its rightful God-glorifying position," and they can be considered to be as official a mouthpiece for creationism as there is.

On their web site (http://www.icr.org) is an essay about the age of the universe. In it, Dr. John Morris, the president of the ICR, says that "every honest attempt to determine the date [of creation], starting with a deep commitment to the inerrancy of God's Word, has calculated a span of just a few thousand years." The Bible thus strongly indicates that the universe is very young. The whole of creation was formed in just six days, it says, and it's possible to get a direct lineal descendency from Adam to historical figures we know existed two thousand years ago. It's difficult to get an *exact* figure from the Bible, but it's clear the number is far, far smaller than the figure physical science would give.

There is a vast amount of data from many different fields of science that indicate the Earth is about 4.55 billion years old. It is not in the scope of this book to detail that information; I instead refer you to any good textbook on astronomy, geology, biology, or physics. My point here is not to show that the scientific view of the age of the Earth and the universe is correct; I will take that as a given. It is the *creationists'* contention that the universe is young that I challenge here. I'll do more than that. I'll say it quite simply: *The young-Earth creationists are wrong.* Utterly and completely.

Creationists usually rely on the Bible for their evidence. They are welcome to believe the Bible is inerrant if they so desire. However, they have lately turned to actual scientific findings to support their claims. But every argument they make is incorrect or incomplete. Every single one. They misinterpret scientific data, willfully or otherwise, and in their writings they only mete out enough information to support their argument, without giving all the data needed to make an informed decision.

I have no intention of discussing their arguments based on the Bible. I leave that to experts on religion and interpreting various ancient texts. I also have no desire to insult, denigrate, or argue against anyone's religious beliefs, *as long as they do not use scientific data incorrectly to support these beliefs.*

Creationists like to say that they practice "creation science." But this name itself is inaccurate; what they do is not science at all. Science is a matter of observation, data, and fact. Religion is a matter of faith and belief. The creationists start with the idea that the Bible is correct, and that any observations of the universe that do not agree with it must be wrong. That is not science; that is dogma.

If some particular scientific observation supports a particular religious belief, that's fine. But when that observation is distorted or misused in some way, that is *not* fine. It is when creationism and science overlap that things get dicey. I won't argue with their belief, but I will happily discuss creationist arguments based on astronomical observations.

The astronomical arguments they use to support their belief in a young universe can be found in their books, religious tracts, and on many web sites. They make dozens of arguments trying to support their untenable position, far more than can be covered in a single chapter of a book. Still, a few of the most common are worth dissecting.

I want you to remember as you read this chapter: these are the challenges the ICR itself and other creationists use against science. These are not straw-man arguments I made up, ideas easy to tear down to make the creationists look bad. These are their *own* weapons, and evidently unbeknownst to them, they are all aimed squarely at the creationists themselves. I have quoted their arguments below, and in some cases have paraphrased them slightly to make them more clear.

> The age of the earth and moon can not be as old as required [by mainstream science due to] the recession rate of the moon (quoted from http://www.icr.org/pubs/imp/imp-110.htm).

A common creationist claim is that the way the Moon's orbit is changing shows that the Moon and the Earth must be very young. Certainly, they say, the Moon is no older than about a billion years, far younger than the age mainstream science declares.

You might think the Moon keeps a constant distance from the Earth, but this turns out not to be the case. Due to the complicated dance of gravity, the Moon's distance from the Earth actually increases by about 4 centimeters (1.6 inches) or so a year (for details, take a look at chapter 7, "The Gravity of the Situation: The Moon and Tides"). This number is well determined, because the Apollo astronauts left reflectors on the lunar surface that can be used by Earthbound astronomers to measure the Moon's distance quite precisely.

If you divide that distance—400,000 kilometers—by the recession rate of 4 centimeters per year, you see it would take the Moon 10 billion years to reach its present distance, assuming it started its journey somewhere near the Earth. However, that naively assumes that the 4-centimeter-per-year rate of recession is constant. Actually, that rate decreases rapidly with distance; the farther the Moon is from the Earth, the more slowly it recedes. In other words, in the distant past the Moon was much closer to the Earth, and receded faster.

If the calculation is performed more carefully, using numbers accounting for this change in the recession rate, you get an age for the Moon of much less than 10 billion years. One creationist, Don DeYoung, found that the Moon can be no older than about 1.5 billion years, and he claims that this is an *upper* limit to the age. According to him, the scientists must be wrong to claim an age of 4.5 billion years for the Moon.

But, again, the creationists are wrong. DeYoung assumed that you could simply extrapolate the Moon's current recession rate backwards in time all the way to when it was formed. As usual in the universe, things are more complicated than that. The current rate is actually much *higher* than usual. The rate depends on how well the Moon interacts gravitationally with the Earth.

The Earth and the Moon interact like a complicated mechanical watch, full of gears; if one slows down, they are all affected. So it is with the Earth and the Moon. The Moon's gravity moves water around on the Earth, causing the tides. This water rubs against the ocean floor, generating friction. That friction takes energy away from the Earth, slowing its rotation, and gives it to the Moon in the form of orbital energy. When the orbital energy of an object is increased, it moves into a higher orbit, so the Moon moves away from the Earth. The increased distance also means the Moon's orbital speed slows.

At this point in history, the Moon's orbit and the Earth's rotation collude to generate a lot of friction with the sea floor, especially near the shorelines of the continents. An unusually large amount of energy is being taken out of the Moon's orbit, causing it to recede faster than it normally would. In a sense, the Moon's gravity has a better grip on the Earth now than it did in the past, and is better at losing its own orbital energy.

What this means is that you cannot say that the current rate of 4 centimeter per year is a good average. In the past, the rate was actually *slower* than this, making the Moon older. DeYoung's estimate of an *upper* limit to the Earth's age actually turns out in reality to be a *lower* limit, and in fact is perfectly in concordance with an age of the Earth and the Moon of 4.5 billion years.

৬ ৬ ৬

Beyond our Earth, the creationists see our very system of planets itself as an indication of the Earth's relative youth.

Astronomers have a pretty good idea about how and when the solar system formed. There have been many theories over the centuries, but repeated observations have indicated that the solar system formed about 4.5 billion years ago (which dovetails nicely with the age of the Earth/Moon system as well). Initially, the solar system started out as a vast cloud of gas and dust. Something caused this cloud to collapse. Perhaps it was a collision with another cloud (which happens fairly often in the Galaxy), a blast from a nearby supernova, or the wind from a red giant star that pushed on the cloud that prompted the collapse.

Whatever the initial cause, as it collapsed, the cloud flattened due to centrifugal force and friction. As the matter in the cloud formed a disk, particles of ice and dust collided, stuck together, and grew. Eventually, over some hundreds of thousands of years, the pieces grew large enough to attract material with their own gravity. When this happened, the disk particles were quickly sucked up by the forming planets. By this time, the Sun was finishing its own formation, and a super-solar wind started. This wind blew any remaining material away, leaving something that looked a lot like the solar system we see today. This theory has recently been strongly supported by many astronomical observations, including those of young solar systems by the Hubble Space Telescope.

The creationists, however, say that the solar system shows several characteristics that are not consistent with the scenario outlined above. The ICR has an educational course available through its web page called "Creation Online," available at http://www.creation online.org/intro\08\8680.htm. In it, ICR officials list several of these arguments. All of these claims are wrong. What follows are verbatim quotations from "Creation Online."

> If the planets and their 63 known moons evolved from the same material, they should have many similarities. After several decades of planetary exploration, this expectation is now recognized as false.

Actually, this claim is false. The disk that formed the solar system was not homogeneous; that is, it wasn't the same throughout its extent. That would be a silly thing for a scientist to assume, since it's clear that near the center of the disk the Sun would heat the material, evaporating off the ice, while near the edge far from the Sun the ice would remain intact.

Astronomers have known for decades that the disk must have had different materials distributed along it, because the outer planets and moons are much different than the inner ones. The outer moons have more ice in their composition, for example, perfectly consistent with a disk that had a distribution of materials along it. To be generous, this argument is at best disingenuous on the part of the ICR. If the collapsed disk theory hadn't jibed with that most basic observation, it would have been thrown out before it ever got proposed.

> Since about 98% of the sun is hydrogen or helium, then Earth, Mars, Venus, and Mercury should have similar compositions. Instead, much less than 1% of these planets is hydrogen or helium.

When they formed, the inner planets probably did have a much higher amount of these gases. However, the gases are very light-weight. Imagine flicking your finger on a small pebble. It goes flying! Now try that on a station wagon. The car won't move noticeably, and you may actually damage your finger. The same sort of process is going on in the Earth's atmosphere. When a molecule of nitrogen, say, smacks into a much smaller hydrogen atom, the hydrogen gets flicked pretty hard, like the pebble. It can actually pick up enough speed to get flung completely off the Earth and out into space. When the nitrogen molecule hits something heavier, like another nitrogen molecule, the second molecule picks up less speed, like the station wagon in our example. It pretty much stays put. After a long time, the lighter atoms and molecules suffer this same fate; they all get flung away from the Earth. Over the lifetime of the Earth, all of the hydrogen and helium in the atmospheres of Earth, Venus, and Mars have basically leaked away, leaving the heavier molecules behind.

Jupiter and the other outer planets retained their lighter elements for two reasons: they are colder, and they are bigger. A colder atmosphere means the collisions occur at slower speeds, so the lighter elements don't get lost to space. A bigger planet also has more gravity, which means the planet can hold on more tightly to its atmosphere. A small hot planet like Earth loses its hydrogen; a big cold one like Jupiter does not.

So the collapsing cloud theory predicts that *initially* the planets may have had a lot of hydrogen and helium in their air, but it's natural—and makes good scientific sense—that some don't anymore.

> All planets should spin in the same direction, but Venus, Uranus, and Pluto all rotate backwards.

According to the collapsing-cloud theory, the planets should all spin in the same direction in which they orbit the Sun because the initial disk spun that way. Anything forming in that disk should spin in the same direction. However, Venus rotates backwards, and Uranus rotates on its side! How can the disk theory explain that?

Actually, the answer is simple: it doesn't. The disk theory concerns only how the planets *formed* and not necessarily how they look *today*. A lot can happen in 4.55 billion years. In this case, collisions happen.

We know for a fact that cosmic collisions occur. We had repeated graphic examples of this in July 1994, when the comet Shoemaker-Levy 9 broke into dozens of pieces and slammed repeatedly into Jupiter, releasing more energy than could humankind's entire nuclear arsenal. Had the comet hit Earth instead of Jupiter, it would have been a catastrophe of, well, *biblical* proportions. Humanity, along with 95% of the land animals on Earth, would almost certainly have been wiped out.

And even this collision is small potatoes. In the early past, when the disk was forming into planets, gravitational interactions would have been common. Two planets forming too closely together would affect each others' orbits, and the smaller one might actually get flung into a wildly different orbit. This orbit could have sent it on a collision course with another planet. An off-center, grazing collision could physically tilt a planet, changing the axis of rotation, in much the same way that poking a spinning top off-center causes the axis to wobble.

In the case of Uranus, a large collision is what most likely knocked it on its side. For poor Venus, whatever collided with it knocked it almost completely heels-over-head. To us it looks like Venus is upside-down and spinning backwards.

Ironically, this catastrophic view of planetary dynamics is more biblical than classically scientific. For many years, scientists avoided using catastrophes to explain events, since they were hard to reproduce, difficult to analyze statistically, and smacked of biblical events. In the end, though, science learned that catastrophes do happen, which is its strength. When presented with evidence contrary to the theory, science learns and grows.

> All 63 moons in the solar system should orbit their planets in the same sense, but at least six have backward orbits. Furthermore, Jupiter, Saturn, and Neptune have moons orbiting in both directions.

This one is really easy to explain. Some of the moons of the planets formed at the same time as the planets and orbit their parent bodies in the "correct" sense, that is, in the same direction that the planet spins and orbits the Sun. However, it's possible, although generally not easy, for a planet to capture moon-sized objects. If the conditions are just right, it's not only possible but rather common that such a captured moon would orbit the planet in the opposite direction. Jupiter and Saturn both have moons that orbit backwards, or **retrograde.** All of these moons orbit at large distances from their planet, as is expected in a capture event as well.

Again, using this as a creationist argument is disingenuous on their part. Retrograde moons have been known about and explained for many decades.

> The sun turns the slowest, the planets the next slowest, and the moons the fastest. But according to evolutionary theories, the opposite should be true. The sun should have 700 times more angular momentum than all the planets combined. Instead, the planets have 50 times more angular momentum than the sun. The sun has 99.9% of the total mass of the solar system, but 99% of the total angular momentum is concentrated in the larger planets.

According to the cloud-collapse theory, the Sun should indeed be spinning faster than any solar-system body. When a figure skater draws in her arms during a spin, she spins faster and faster. The

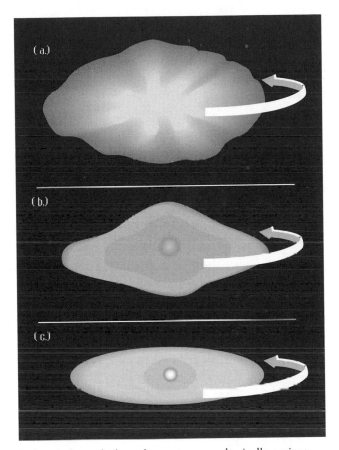

Before it formed, the solar system was basically a giant cloud of dust and gas, roughly spherical in shape. As it collapsed, it began to spin, and this caused it to flatten out. The planets formed *after* the collapse, which is why they all orbit the Sun in roughly the same plane.

fancy name for this is **conservation of angular momentum,** which just means that a big thing will spin faster if it shrinks.

This happened to our proto–solar-system cloud when it collapsed, too. It spun faster as it collapsed. Since the Sun was at the center of the cloud, it should have sped up the most. But, apparently contrary to this theory, the Sun currently spins only once a month. This is the key to the last of these creationist arguments about the solar system.

The creationists, as usual, are oversimplifying the problem. The universe is many things, but one thing it isn't is simple. However, it usually makes sense, so if you see something that doesn't make sense, look around. The solution may be blowing right past you.

In this case, that's literally true. The Sun is constantly blowing a wind. This **solar wind** flows from the Sun's surface at a rate of about a thousand tons of gas per second. But the Sun is pretty big and has mass to lose. It can easily shed a few billion tons a year and hardly notice the difference.

The solar wind is made up of charged particles—electrons and protons. Left to themselves, they would blow straight out from the Sun into infinity. But they are not left to themselves. The Sun itself has a pretty hefty magnetic field, and this field rotates with the Sun. As it spins, the sun drags the particles along with it. This in turn acts like a brake on the Sun's rotation, slowing its spin.

This isn't too difficult to understand. Imagine standing in your front yard with a big trash bag in your hands. Open it up, hold it out to your side, and start spinning. The bag acts like a parachute, scooping up air and slowing you down. The exact same thing is happening to the Sun. The magnetic field is like a huge parachute, scooping up particles. The "air"—the particles in the solar wind— is very tenuous, and the Sun is large and heavy, but this drag (despite what the creationists might think) has had a long time in which to work. Over 4.5 billion years it has quite possibly slowed the Sun substantially to its current rather stately monthly rotation. While this theory has not been conclusively proven, it remains a leading contender to explain the angular momentum problem. There are other theories as well, such as the idea that the Sun lost most of its angular momentum very early on, as a protostar. It may have shed a lot of its mass through long episodes of a sort of super-solar wind. While astronomers are not totally sure which of these ideas is the correct solution, the fact remains that there *are* plenty of ideas, and they use good, solid physics.

There are not enough old supernovas to justify an old galaxy.

Of all the creationist arguments involving astronomy, this one is my favorite. Basically, it goes like this: some stars, at the ends of their lives, explode. This doesn't happen terribly often, and only stars far more massive than the Sun ever explode in this way. When they do, it's called a **supernova.**

The explosion is so bright that it can outshine entire galaxies and be seen clear across the universe, and it is so violent that the outer layers of the star are flung outward at an appreciable fraction of the speed of light. This rapidly expanding cloud of debris, called the **ejecta,** or sometimes the **remnant,** can glow for many tens or even hundreds of thousands of years.

You can see where this is going. Creationists take exception to the idea of old supernova remnants, of course, since according to them, none can be older than 6,000 years. As a matter of fact, a relatively new and bold claim by the creationists is that there is no supernova remnant older than at most 10,000 years. This argument has been used by noted creationist Keith Davies, and is also used by the Institute for Creation Research itself. They list it prominently at their web site, and it carries a lot of their weight. If it's wrong, then so are they.

It's wrong. Davies goes through a tremendous number of observations and calculations to show that, according to mainstream science, there should be lots of very old supernova remnants in the sky, yet none is to be found. He uses lots of math and fancy graphics to prove his point.

The funny thing is that he is missing the forest for the trees, so to speak. I could go into just as much detail showing why there really *are* supernova remnants older than 6,000 years, and some that are actually hundreds of thousands of years old. But I don't need to. Even if we grant that there are no remnants older than 6,000 years, it doesn't matter. Davies's whole line of attack is wrong for a very simple reason: *Supernova remnants were not created at the moment the universe began.*

According to the creationists, this (supposed) lack of old remnants indicates the universe is young. But remember, the remnants form when a star goes supernova *at the end of its life.* Stars live longer than 6,000 years—far, far longer. The absolute minimum age at which a star can blow up is roughly a million years old, so no matter how you slice it, the universe must be *at least* that old for us to see supernova remnants at all. So, even if we grant that the oldest remnant is 6,000 years old, the universe must be 1,006,000 years old at least for us to see it.

That's why this is one of my favorite creationist arguments. In this case, they are using a sort of sleight-of-hand, a magician's trick to distract you by using complex mathematics, when in fact the argument rests on a fatally flawed premise. You don't need to do any fancy math at all; a little logic destroys their argument.

Incidentally, as an indication of Davies's misunderstanding of all this, he has an image of a star on his web page that he has labeled as a supernova. It isn't. It's really just a plain old star that's been overexposed, a fact that is easy to verify by simply opening nearly any astronomy textbook. Ironically, the very first thing you see on his (very long) web page shows that he doesn't understand what a supernova is at all.

❧ ❧ ❧

The creationists' attack on science is a serious issue. It goes far beyond bad astronomy. Indeed, astronomy is only the most recent of repeated attacks on mainstream science that they have initiated. Their feelings about biology are well-known in the United States. In 1999 the Kansas School Board discouraged the teaching of biological evolution in middle and high schools by removing all statewide standardized testing about it. This was accomplished because the school board had been packed with creationists in the previous election. Before the election, the creationist candidates had downplayed their religious ties. They also relied on the voters being too apathetic to research their candidates' histories. That gamble paid off, and the result was a creationist school board, a nationwide controversy, and a terrible embarrassment for the Kansas—and American—educational system.

Perhaps more frightening were the reactions of other politicians to this. Several contenders in the 2000 presidential election were sympathetic to the board's decision, without any understanding of the lack of scientific reasoning behind it.

One of my favorite phrases from the Bible is, "The truth shall set you free." Ironically, creationists don't want you to know the truth. They want you to know what *they say* is the truth, and science isn't like that. Scientists understand that the universe is trying to *show* us the truth, and all we need to do is figure it out. It's clear from this short selection of astronomical topics that when it comes to science and critical thinking, creationists are selling a bill of goods. My advice: don't buy it.

A happy addendum: upon the next election, the public having learned of their true agenda, three of the creationist Kansas School Board members were voted out of office. One spent an unprecedented amount of money in advertising and was ousted, anyway. The new board quickly reinstated teaching evolution in the curriculum. Like science, sometimes even the political process is self-correcting.

20

🙙🙙🙙🙙🙙🙙

Misidentified Flying Objects: UFOs and Illusions of the Mind and Eye

O n February 11, 1997, at approximately 3:00 A.M. local time, I had a close encounter with a UFO. Actually, multiple UFOs.

I was in Florida with my family to attend a Space Shuttle launch. I had been working at Goddard Space Flight Center in Maryland for nearly two years, helping to calibrate a new camera that was to be placed on board the Hubble Space Telescope. All of us who had worked on the camera got passes to see the Shuttle launch in Florida, and we were all excited about seeing our camera lofted into space.

The launch was scheduled for 3:55 A.M. That's not the best time to appreciate the spectacle and fury of rocket launch, but the vagaries of orbital mechanics demanded such a liftoff. My mom volunteered to baby-sit our infant daughter Zoe, so my father, my wife, my nephew, and I made our way to Cape Canaveral around 1:00 A.M. We quickly found out that several thousand other people were also attending. Too excited to nap, my father and I wandered around talking to the other attendees. Many people had telescopes set up to watch the distant shuttle, proudly standing under the intense glare of multiple spotlights. To see it we had to peer across a long stretch of the Banana River that separates the cape from mainland Florida. The cape is surrounded by water, and by wildlife. We actually saw a couple of alligators, which is a weird sight so close to such a technological marvel.

About an hour before the launch, I spotted some unusual lights in the night sky, a dozen or more, to the right of the launch pad from our viewpoint. They were perhaps at the same distance from us as the pad, about 10 kilometers (6 miles), although it was hard to tell. My father pointed out that they were moving, so we kept watching. The movement was very slow, as if they were hovering. I figured it was a group of distant spotting planes, but then remembered that NASA only uses one or two planes to sweep around the area. No other planes are allowed near the shuttle; it is jealously guarded by NASA for obvious reasons.

My next guess was birds, but these objects were *glowing*. Balloons? No, they were moving too quickly. No satellites group together like that. My excitement mounted, despite my more rational thoughts. What were they? As I watched, I noticed that they were moving together, but *not* in a straight line. They weaved slightly. That ruled out satellites and a host of other mechanical objects.

I refused to think of any ridiculous explanations involving anything, well, ridiculous. But what were these things? All I could see through binoculars were glowing dots.

Their flight path was taking them to my right as I continued to watch them through binoculars. Slowly, faintly, I could hear a noise they were making. It was eerie, odd, difficult to place. Then, suddenly, the noise got louder, and the objects in my binoculars resolved themselves. My mind and heart raced. I was seeing . . .

. . . a flock of ducks. As they flew by us they were just a few hundred meters away, and they were unmistakably terrestrial waterfowl. The noise we heard earlier was their quacking, muffled by distance, and their otherworldly glow was just the reflected light of the fleet of spotlights flooding the Shuttle pad. The ducks' weaving flight was obvious now, too. They appeared to be hovering when we first saw them because they were so far away and were heading roughly toward us.

I never allowed myself to think that they were truly UFOs, but what was that odd feeling in the pit of my stomach while I watched them, and why was I vaguely disappointed when we identified them as ducks? I laughed to my dad, maybe a little bit too loudly, and we resumed our vigil over the shuttle.

❧ ❧ ❧

I learned two interesting lessons from this experience. Well, three: the first would be not to mistake ducks for alien spaceships. But the other two are a bit more profound. One is that there is a human need to believe in extraordinary things. Over the course of our lives we build a mental database of ordinary events. We see trees, airplanes, buildings, people we know, and we catalog them in our minds. When we see something that doesn't fit into the picture we have of life, it can be hard to categorize. It's easy to get excited by it, to wonder about it. Sometimes we wind up either identifying it as something we already know or setting up a new category for it.

This happens in science all the time. Say a scientist spots a new phenomenon. It might turn out to be something we already know about that is seen in a new way, or maybe it's something actually new that deserves study. But so far, with all the observations made by thousands or even millions of scientists, not a single phenomenon has ever been shown to be anything but natural, and certainly nothing appears to be guided by an intelligent hand not our own.

But the need to *believe* in such things is firmly planted in our collective psyche. There is something wondrous in seeing something we cannot explain. I like mysteries, for example, and I'll worry over them until I can solve them. I think there may be some hardwiring in our brains that almost demands us to want mystery in life. If everything were explained, where would the fun be? So even I, a hardheaded and skeptical scientist, once allowed myself to be momentarily swept up in a nonrational thought process.

The third lesson from my close encounter that night at Cape Canaveral is that even an astronomer with years of experience and training in identifying objects in the sky can make a mistake, even a silly one. Together with a series of unusual circumstances (the objects were glowing, they were distant, they were headed roughly toward me) it was possible and perhaps even easy to make a false conclusion, or at least skip over the correct one.

In my opinion, these two processes—a need for wonder, and an all-too-easy ability to be fooled—account for the vast majority of UFO sightings. I am not a social psychologist, so I don't want to ponder any further about the human desire to wonder. It's fun to think about it, but I am not qualified to analyze it other than as a layman.

But I *am* a scientist and an astronomer. So let's take a look at the visual and physical phenomenon of UFOs.

A lot of people claim to see strange things in the sky—moving lights, changing colors, objects that follow them. But let's think about this for a minute: how many people are really familiar with the sky? I have found that there are things that happen in the sky about which people are completely unaware. Many have no idea you can see planets and satellites with the naked eye. When I talk to the public about such optical phenomena as haloes around the Sun and **sundogs** (teardrop-shaped glowing patches in the sky that are caused by sunlight being bent by ice crystals in the air), the vast majority has never heard of them, let alone seen them.

If someone is not familiar with things that are in the sky all the time, how can they be sure they are seeing something unusual?

This is not as ridiculous as it sounds. Venus, for example, can be amazingly bright, far brighter than any starlike object in the sky—in fact, it's the third brightest object in the sky after the Sun and the Moon. Seen low on the horizon, Venus can flicker in brightness and change color as the atmosphere bends its light. If you are driving, it can appear to follow you through the trees.

It's common to mistake brightness for size. I receive e-mails from people all the time asking me about the huge object they saw in the sky, and it usually turns out to be Venus. This planet is so far away that to the unaided eye it looks like a star, although a bright one. Unless you have unusually sharp vision, without at least a pair of binoculars you can't see Venus as anything other than an unresolved dot. Yet, because of its tremendous radiance, Venus is commonly mistaken by people as a huge disk in the sky. That's why it accounts for the majority of UFO sightings.

If you are not familiar with the sky, any unusual object can make you jump to erroneous conclusions. But there is a group of people who are very familiar with the sky—by some estimates there are as many as 100,000 amateur astronomers in the United States alone—who spend many hours every week doing nothing but looking into the sky. They own telescopes and binoculars and spend every clear night outside looking up.

Think about that for a moment. These folks are looking at the sky *all the time*. Yet, of all the people I have had approach me or e-mail me to say they have seen a UFO, not one has been an amateur astronomer. As a matter of fact, I have never heard about any amateur astronomers seeing something in the sky they absolutely could not explain. Yet they spend far more time looking at the sky than lay people and statistically should see far more UFOs! How can this be?

Easy. Remember, the amateur astronomers study the sky. They know what's in it and what to expect. When they see a meteor, or Venus, or sunlight glinting off the solar panel of a satellite, they know it's not an alien spaceship. Amateur astronomers know better and, in fact, all the amateurs to whom I have spoken about this are very skeptical about UFOs being alien spaceships. This is a very strong argument that there are mundane explanations for the vast majority of UFO sightings.

❧ ❧ ❧

Eyewitness accounts are notoriously unreliable. However, UFO enthusiasts usually point out that we have far more evidence than these accounts. We have cameras.

We have all seen footage of UFOs on television. Usually it's an amateur photographer, perhaps someone on vacation with a video camera, who sees a distant object and quickly gets it on tape.

I am always immediately suspicious of such sightings, specifically because of my own experience with the ducks. A blurry object seen from a great distance is a poor piece of evidence for extraterrestrial visitors. It could be any number of common things, from ducks to balloons. Airplanes headed straight at you can appear to hover for a long time (I once thought an airplane making an approach to an airport behind me was the planet Mars; it was stationary in the sky and glowing red). Helicopters actually *can* hover, and have odd lights on them. A shakily held camera makes the object appear to move. I have seen quite a few TV shows showing footage of a breathless videographer exclaiming how an object is moving, when it's obviously the unsteady hand of the person that is moving the camera.

Worse, the camera itself distorts the image. A famous UFO tape shows a faint dot that, as the camera zooms in, gets resolved as a diamond-shaped craft. Actually, the diamond shape was due to the internal mechanisms of the camera, and when the videographer zoomed in the object took this shape because it was out of focus.

Modern electronic cameras have all sorts of odd defects that can distort images in unfamiliar ways. Another famous series of shots shows UFOs that are very bright with a very dark spot trailing them. UFO believers claim that this is due to some sort of space drive using new physics we don't understand. Actually, this is more likely an effect in the camera's electronics. A bright object can cause a dark spot to appear next to it in the camera's detector due to the way the image is generated. I have seen similar effects in Hubble Space Telescope images.

My point is, don't attribute to spaceships what you can attribute to yourself or your equipment.

On the pseudodocumentary TV show *Sightings,* which gullibly and unskeptically presents all manner of pseudoscience as fact, I saw a segment in which a photographer claims that hundreds of UFOs can be seen all the time. He puts his camera directly underneath an awning and points it to just below the Sun. The awning shades the Sun just enough to put the camera in shadow. He then turns it on, and voilà! You can see dozens of airborne objects flitting this way and that. He calls this method the "solar obliteration" technique, and says that without it we would never see the flying objects.

The photographer claimed that these were UFOs. I was amazed; he had nothing on tape but fluff blowing in the wind. He didn't bother performing even the simplest of tests to try to find out what these things were. If they were cottonwood seeds, for example (which is what they look like to me), a fan blowing near his camera might settle the issue. If that didn't work, he could set up two cameras spaced a few meters apart and aimed in the same direction. A distant object would appear in slightly different places in the cameras' fields of view. You can see this yourself by holding your thumb up at arm's length and looking at it first with one eye closed, and then the other. Your thumb will appear to move back and forth compared to more distant objects because the angle at which you are viewing it is changing. This method, called **parallax,** can be used to determine the distance to objects. Our intrepid UFO photographer never tried it, so we may never know if his objects were interstellar travelers or simply a tree trying to be fruitful and multiply.

Another thing to note when listening to UFO claims is the estimation of size and distance. This is always a red flag for me when I hear a UFO report. Someone says it is a kilometer across and twenty kilometers away, but how do they know it isn't a *meter* across, and twenty *meters* away? Without one measurement, you cannot possibly determine the other.

People who should know better make this mistake as well. Take Dr. Jack Kasher, for example. He is a physics teacher at the University of Nebraska at Omaha. He believes that UFOs are in fact spaceships populated by aliens. His claim to fame involves a bit of footage from STS 48, a Space Shuttle flight from 1991. During the mission the cameras were pointed down, toward the Earth. The cameras used at night are extremely sensitive and can see outlines of the continents, even in the dark.

In the now-famous footage, we see specks of light moving in the camera's field of view. Suddenly, there is a flash of light. One of the specks then makes a sharp-angled turn, and another shoots in from off-camera, going right through where the other speck was.

Kasher claims that this is evidence of alien spacecraft. The first point of light is an alien ship. The burst of light seen is the flare from a ground-based missile launch or a secret test of a Star Wars defense. The second point of light is the missile or beam weapon itself. The first dot, the alien ship, then makes an evasive maneuver to avoid being blown back to wherever it came from. According to Kasher, the film has captured an interplanetary battle.

Needless to say, I disagree with him.

So do a lot of other folks. These include Shuttle astronaut Ron Parise and space program analyst James Oberg. Both have discussed what *really* happened on STS 48. The specks of light are actually bits of ice floating near the Shuttle. The particles of ice form on the outside of the Shuttle on every mission, and can get jolted loose when the rockets fire. Once separated, they tend to float near the Shuttle. The flash of light seen was a vernier rocket, a small rocket that controls the direction in which the Shuttle points. It does not generate much thrust, which is why you don't suddenly see the Shuttle moving during the burst. (Kasher claimed that a rocket firing would obviously move the Shuttle but neglected to research just how much thrust the rocket gave off.) The rocket burst hitting the first bit of ice is what suddenly changes its course, and the second bit of light flashing by is simply another ice particle accelerated by the rocket. If you look at the footage closely, you can see it doesn't actually get very close to the first particle, making this a poor demonstration of Star Wars technology.

Kasher has made quite an industry of going on TV shows and showing this footage, which he clearly does not understand. He even sells a video of his analysis of the footage ($29.95 plus shipping and handling). I'd save my money if I were you.

🍃 🍃 🍃

I am commonly asked if I believe in life in space, and if aliens are visiting us. I always answer, "Yes, and no." This confuses people. How can I believe in aliens when I don't believe in UFOs?

It's actually easy. Space is vast, terribly vast. There are *hundreds of billions* of stars in our Galaxy, and it's becoming clear that many—if not most—have planets. There are billions of galaxies like ours in the universe. In my opinion, it's silly to think that in all the universe we are the only planet to have the right conditions for life to arise.

But even if the Galaxy is humming with life, don't expect ET to come here to poke at us, draw funny patterns in our corn fields, and mutilate our cattle. The very vastness of space makes that unlikely. Even with highly advanced technology, it would be a lot of work to explore every star and planet in the Milky Way. And if their technology is so advanced, how come they crashed here in Roswell in 1947? It seems unlikely that we would be able to shoot down a spaceship; that's like cows being able to take down a fighter plane. And if their technology is so advanced, why would they crash a few kilometers from the end of a journey that lasted for trillions of kilometers?

Many people assume that faster-than-light travel is possible. Although there is no real evidence for it, let's assume that travel through hyperspace, warpdrive, or some other method is feasible.

If that's true, and some alien race knows about it, where are they?

Our civilization has been around for a few thousand years, and we are starting to explore the near reaches of space. If we had faster-than-light travel, we could populate the entire Galaxy in just a few thousand years. Even with slower-than-light ships we could do it in a few million years at most.

That may sound like a long time, but not in the galactic sense. Elsewhere, a star like the Sun, only a few hundred million years older, might have had a booming civilization on one of its planets while trilobites swam in our own oceans. If this civilization decided to colonize the Galaxy, then by now the whole place should be filled with them.

Yet we have no proof they are here, so you have to assume that they have some sort of Prime Directive, as in *Star Trek*, to leave young civilizations alone. But then, why do we see so many of their spaceships? UFO enthusiasts want to have it both ways; they believe that incredibly advanced aliens come here in spaceships, yet these aliens are still so dumb that a bunch of primitives who barely understand atomic energy can photograph them ten times a day. That stretches credulity beyond the breaking point. I think the Truth Is Out There. It just isn't down here.

It frustrates me to see true believers claim that every light in the sky is an alien spaceship. However, it gives me pause when I remember that I was once fooled by a flock of ducks on a dark night at the most rational of locations, a rocket launch. It helps me remember that anyone can get fooled; I just wish that more people would be more critical about what they see.

As for that Shuttle launch, it was successful despite any fowl play. The camera was installed perfectly, and returned many gigabytes of interesting and useful information about what is really going on in space. And I will admit it here, in this book, for the first time: in some of those images I *did* see evidence of intelligent life in space. In many images from Hubble I have seen long, bright streaks of light that were clearly not cosmic ray tracks, misguided tracking, asteroids, comets, or moons.

What were they, then? They weren't alien spaceships. The long streaks were caused by human-built satellites, placed in orbits higher than Hubble. As they passed through Hubble's field of view, their motion left a streak of light.

We have met life in space, and it is us.

21

Mars Is in the Seventh House, but Venus Has Left the Building: Why Astrology Doesn't Work

"The fault lies not in our stars, but in ourselves."
—*Julius Caesar* by William Shakespeare

People say the weirdest things.

A while back, my sister threw herself a nice birthday party. She invited her friends, her family, and her coworkers, and we all had a great time. She had a special setup for kids to play and do artwork, and while I was chatting with some of the other party-goers my daughter ran up to me to show me pictures she had drawn. One drawing in particular caught my eye. I said, "Hey, that looks like an object I study at work!"

A woman next to me asked, "What is it you do?"

"I'm an astronomer," I replied.

Her eyes got wide and her face lit up. "That's so *cool!*" she said. "I'm terrible at astronomy. I failed it in college."

Well, what do you say to *that?* I let it slide, and we talked for a while about this and that. A few minutes later, she was commenting on one of the people at the party, and said, "Oh, he's a Libra; all Libras do that."

Ah! *Now* I knew why she failed astronomy. I refrained from saying anything out loud about astrology, though. I knew better.

The basic premise of astrology is simple: the arrangement of the stars and planets at the time of our birth affects our lives. There is no evidence to show that this is true, but people believe it nonetheless. A 1984 Gallup poll showed that 55 percent of American teenagers believe in astrology. Clearly, astrology is popular.

But that doesn't make it right.

Astrology is like a religion to its devotees. Many religions require faith without proof. In some sense, proof is anathema to these beliefs; you are supposed to believe without needing to see any evidence. Astrology is the same way. If someone tries to show just exactly why astrology doesn't work, you run the risk of being tarred and feathered.

Luckily, I'm willing to take that risk, and I'll say it here, unequivocally and without hesitation: astrology doesn't work. It's hooey, hogwash, and balderdash.

Astrology lacks any sort of self-consistency in its history and in its implementation. There is no connection between *what* it predicts and *why* it predicts it, and, indeed, it appears to have added all sorts of random ideas to its ideology over the years without any sort of test of the accuracy of these ideas.

Science is the exact opposite of this. Scientists look for causes and use them to make specific predictions about future events. If the theory fails, it either gets modified and retested or it gets junked. I'll note that science has been spectacularly successful in helping us understand our universe, and that it is perhaps the most successful endeavor undertaken by humans. Science works.

Astrology, on the other hand, doesn't. It makes vague predictions that can always be adapted after the fact to fit observations, as we'll see. Astrologers don't seek causes at all, for a good reason: *There isn't any cause to astrology.* If you look for some underlying reason, some connection between the stars and planets and our lives, you won't find any. For astrology to sell, buyers must not seek out the fundamental principles behind it, because if they do they'll see that there is none.

To avoid making testable claims about the driving forces of astrology, astrologers rely on mumbo-jumbo to bamboozle the public. If in the rare case they actually resort to specific arguments to defend astrology, they frequently use misleading terms. One web site (www.astrology.com) defines the basic force behind astrology by stating:

> In some ways, the forces between the Planets involved in Astrology can be simplified into one word: gravity. The Sun has the greatest gravity and the strongest effect in Astrology, followed by the Moon, the Earth's satellite. The other Planets are not truly satellites of the Earth, but nevertheless, they have gravity and so affect the Earth. The Sun controls the Earth's motion and the Moon controls its tides, but the other Planets have their own effects on the Earth—and on the people who live on Earth. Sometimes their influences can be so strong that they outweigh the Sun's energy!

But this makes no sense at all. If gravity is the dominant force, why doesn't it matter if Mars is on the same side of the Sun as the Earth when you are born, or the opposite side? Mars' gravitational influence on the Earth drops by a factor of more than 50 from one side of the Sun to the other. One would think this would be an incredibly important detail, yet it is ignored in most horoscopes.

It's also easy to show that the Moon has a gravitational effect on the Earth (and you) that is more than 50 times the combined gravity of the planets. It seems to me that if gravity were the over-riding factor in astrology, the Moon would influence us 50 times more than do the planets.

Astrology apologists sometimes look to other forces like electromagnetism. That's an even worse choice than gravity, since the Sun's effect on us is millions of times that of any other object in the sky. The Sun's solar wind of charged particles is what causes aurorae, and a strong electromagnetic outburst from the Sun can trigger electricity blackouts and even damage satellites. This is a genuine physical effect and will have far more influence on your day than any horoscope.

Astrologers must then rely on some force that is not like gravity or electromagnetism. Some claim this force does not decrease with distance, thus sidestepping the problem of the planets' true distance when you are born. But this opens up a new can of worms:

as I write this, more than 70 planets have been discovered orbiting other stars. If the force behind astrology does not decrease with distance, what do we make of those planets? How do they affect my horoscope? And there are hundreds of billions of stars in our Galaxy alone. If they have as much force as, say, Jupiter and Mars, how can anyone possibly cast an accurate horoscope?

Not to mention other bodies in our own solar system. The discovery of Uranus, Neptune, and Pluto threw astrologers into a fit for a while, but they were able to subsume these planets into their philosophy. Interestingly, one web site even mentions the four biggest asteroids: Ceres, Vesta, Pallas, and Juno. These are named after female gods, and the web site gives them feminine attributes after the goddesses for which they are named. Ceres, for example, was the goddess of fertility and the harvest, and astrologically (so claims the web site) has power over a woman's procreative cycle.

But Ceres was discovered in 1801 and named by its discoverer, Giuseppe Piazzi, who happened to choose a female name. Traditionally, all asteroids are now named after women. So are we to believe that an object named rather randomly by a man, and a series of objects named traditionally after women, really have the aspects of the goddesses for whom they are named? What do we do with asteroids like Zappafrank? Or Starr, Lennon, Harrison, and McCartney? My good friend Dan Durda has an asteroid named after him. I don't have any idea how asteroid 6141 Durda (as it is officially called) affects my horoscope personally. If it collides with another asteroid and breaks apart, should I send flowers to Dan's family?

Despite the claims of its practitioners, astrology is not a science. But then what is it? It's tempting to classify it as willful fantasy, but there may be a more specific answer: magic. Lawrence E. Jerome, in his essay "Astrology: Science or Magic," makes a strong claim that astrology is more like magic than anything else (*The Humanist* 35, no. 5 [September/October 1975]). His basic assertion is that astrology is based on the "principle of correspondence," the idea that an object has some sort of effect on reality by analogy, not by physical cause. In other words, Mars, being red, is associated with blood, danger, and war. There is no physical association there, just analogy. This is how magic works; sorcerers take an object like a doll that, to them, *is* what it represents, like an enemy king. Anything they do to the doll happens to the king.

Despite our deepest wishes, though, the universe doesn't work on analogy (would *Star Trek*'s Mr. Spock's green blood indicate that blue-green Uranus is the Vulcan god of war, I wonder?). The universe works on *physical* relationships: The Moon's gravity affects tides on Earth, nuclear fusion in the Sun's core eventually heats the Earth, water expands when it freezes because of the geometry of ice crystals. All of these events pass the test for being real: They have consistant physical rules behind them, they are able to be modeled using mathematics, and these models can be relied upon to accurately predict future events. Also, these events are not subject to interpretation from one person to another.

Astrology isn't like that. The color of Mars may look blood red to one person, but it looks rusty to me (and indeed, the surface of Mars is high in iron oxide—rust). Maybe Mars should represent decay and age, like an automobile in a rainy junkyard, and not the martial aspect of war. Astrological correspondence is up for grabs depending on who uses it. It's not consistent, and it fails the other tests as well.

The shapes of the constellations are another indication of astrology's failures. Technically I am a Libra, having been born in late September. Countless horoscopes tell me that this is the symbol of balance and harmony. Yet look at the constellation: it's basically four rather dim stars in a diamond shape. You can perhaps imagine a set of old-fashioned scales there, implying balance. But it looks more like a kite to me. Should I then be lofty, or an airhead, or prone to windy proclamations (hmmm, don't answer that)? To modern eyes, the constellation Sagittarius looks only vaguely like an archer, but far more like a teapot. The Milky Way is thick and dense in that area of the sky, looking for all the world like steam rising from the spout. Do people born under that sign quietly boil until they explode into a heated argument? The constellation Cancer has no stars brighter than fourth magnitude, making it difficult to see from even mildly light-polluted skies. Are Cancers quiet, faint, dim? Why should the ancient Arabic or Greek constellations be any more valid than mine?

Mind you, the shapes of these constellations are arbitrary as well. Libra looks like a diamond only because of where we are relative to those four stars. Those stars are at all different distances, and only appear to be a diamond. If we had a three-dimensional view, they wouldn't look to be in that shape at all.

And it gets worse. Some stars in the zodiacal signs are supergiants and will someday explode. Antares, the red heart of Scorpius, is one of these supergiants. Someday it will become a supernova, and Scorpius will be left with a hole in its chest. How do we interpret the constellation then?

ɞ ɞ ɞ

Apologists for astrology, like many who defend a pseudoscience, try to distract critics rather than actually argue relevant points. Many astrologers point out that astronomy and astrology used to be the same thing, as if once having been part of a physical science legitimizes astrology. That's silly. That hamburger I ate the other day was once part of a cow; that doesn't make me a four-legged ruminant, and it doesn't make the cow any more human.

Another classic astrology defense argues that many famous astronomers were practicing astrologers: Kepler, Brahe, Copernicus. Notice that the list features astronomers from a few hundred years ago. In the end, this argument is just as fallacious as the previous one. Astronomers from centuries past didn't have the scientific basis for astronomy as a physical science that we now have, and, indeed, Kepler was the key person in making that happen. They were still steeped in tradition. Also, it's not clear if Kepler believed in astrology; he was being paid by a king who did, and he was certainly smart enough to understand who buttered his bread.

Astrologers go on to talk about the large number of people who believe in and practice it. Is the majority always right? Fact is fact, unswayed by how many people believe in a falsehood or how fervently they defend it.

Yet astrology is still popular, despite all these devastating claims against it. Why? What weapon do astrologers wield that wipes out all rational and critical claims against them? It turns out their best weapon is us.

When people read their horoscopes, they tend to report an uncanny number of "hits" or correct guesses. How many of us have read a horoscope and been amazed at how well it described our day?

As an experiment, I put my own birth date into a web-based horoscope generator. It reported several things that do indeed describe me: I like to avoid conflicts (despite the tone of this and other sections of this book), I seek a partner who is my intellectual equal, and I prefer to be with other people over being alone. All true. But it also read: "You are a gentle, sensitive person with a deep understanding of people and a very tolerant, accepting, non-judgmental approach toward life." My wife (who is *at least* my intellectual equal) nearly split her side laughing when she read that.

But let's take a look at those apparent hits. The description above sounds like a lot of people I know and not just me. The wording is vague enough to apply to just about anybody. This is the basic methodology of the astrologer: wording that applies to everyone. People will pick and choose the parts they want to remember, and that is what reinforces the belief in astrology.

The well-known skeptic and rational thinker James Randi (better known as The Amazing Randi) once performed an experiment in a schoolroom. The teacher told the class that Randi was a famous astrologer with an incredible record of accuracy. In advance the teacher had the students write down their birth dates and place each in a separate envelope. Randi cast a horoscope for each person in the room, placing them in the corresponding envelopes, which were then handed back to the students.

After the students read their horoscopes, Randi polled them about accuracy. The majority of the students thought the horoscopes cast for them were accurate, and very few said they were inaccurate.

But then Randi did an Amazing thing: he asked the students to hand their horoscope to the person sitting behind them (the students in the last row brought theirs up to the front row), and then read their neighbor's horoscope.

The results were priceless. Surprise! Randi had put *the exact same horoscope in each envelope.* You can imagine the expressions of shock, then chagrin, then embarrassment that crept over the faces of the students. The wording Randi used was vague enough that it applied to nearly every student in the room. He used phrases like "You wish you were smarter," and "you seek the attention of others." Who doesn't?

A specific horoscope might be wrong. A vague one never is, which is why horoscopes are generally very vague indeed. The complementary aspect is the all-too-human ability to forget bad guesses and remember accurate ones. Astrologers rely on our ability to forget the misses in order to continue bilking millions of dollars from the public.

And bilk they do. Astrology is a vast business. Perhaps most appalling is the appearance of a horoscope in daily newspapers across the country. In their defense, the newspaper editors claim they don't believe in it, either, and place the horoscopes in the comics section, indicating how seriously horoscopes are taken. But that's a cheat: the comics are one of the most popular sections of the paper, and the horoscopes are there to increase visibility, not to take away credibility. If the editors don't believe them, why are they there in the first place?

One of the biggest pro-space web sites is space.com. They have a huge array of pages devoted to space news, history, opinions, and anything you can think of related to space travel. One day some of the business people in charge decided it would be a bright idea to have horoscopes on the site. They put them up, and within days (or more likely hours) received so many angry e-mails from people protesting the horoscopes that the decision was hastily made to take those pages down. I have no doubt that there was some disconnect between the business end of the site and the content folks (I can imagine the business people figuring astrology has something to do with stars, and that's space-related, right?), but in the end the correct thing was done, and hopefully a lesson was learned. I wish the same could be said about newspapers.

On second thought, maybe I was wrong a moment ago: putting horoscopes in newspapers is not the most appalling aspect here. I think the most disturbing part is the pervasiveness of astrology. It's a numbers game; as long as enough people are fooled, it can be self-sustaining. Stories are told, critical thinking goes out the window, and more people believe. Where does it stop? People laughed when former U.S. president Ronald Reagan's wife Nancy relied on an astrologer to make appointments for meetings when the signs were auspicious, but this is no laughing matter. He was *the president of the United States,* and *his wife was relying on an astrologer!* That's just about the scariest thing I can think of. I would hope someone wielding that much power would have just a wee bit more rational thinking ability.

❧ ❧ ❧

Incidentally, the horoscope cast for me on the web did have more to say, and it sums up this chapter better than anything I could possibly write:

> Though you may be as intelligent as anyone, you do not really have a rational, logical approach toward life, and trying to reach you through logical arguments is often futile. Your feelings, intuition, and heart, not your head, lead you, which may infuriate or bewilder your more rational friends. You certainly recognize that there is much more to life than can be explained intellectually and categorized into neat little boxes, and you have an open, receptive attitude toward such areas as psychic phenomena, telepathy, parapsychology, etc.

After reading this chapter, wouldn't you say that sounds just like me?

PART V

✇ ✇ ✇ ✇ ✇ ✇

Beam Me Up

We've traveled a long way in this book, from rooms in which we live to the ends of the universe and finally back again, to plunge into the cobwebbed depths of the human mind. Where, of course, there is still more bad astronomy. Don't think that the next chapters are in this last section because the topics didn't fit anywhere else in the book! Oh no, *these* chapters are special, so special that I decided that they didn't belong earlier in the book. (That, and they didn't really fit in with anything either.)

In this final section we'll see the best and the worst of ourselves as reflected in our works. We'll start with the Hubble Space Telescope, certainly the most abused $6 billion observatory ever built. There are so many misconceptions about Hubble that a whole book could be written about them. I hope you'll settle for just a single chapter.

Hubble may have cost a lot, but you don't need the gross national product of a country to buy the stars. Some companies will sell stars to you for a substantially reduced, though not inconsiderable price. But this depends on what you mean by "sell." It's not really stars these companies sell so much as a bill of goods. They promise the sky, but all they deliver is a piece of paper of dubious astronomical value. And there are darker implications of this transaction, too.

Our final exploration of bad astronomy takes us back to the silver screen. Epic myths may be Hollywood's biggest staple, but Tinseltown's grasp of science has never been the best. The box office may be the biggest purveyor of bad astronomy that exists. Spaceships don't roar through the skies, asteroids aren't that big of a danger, and aliens aren't likely to take a pit stop at Earth long enough to eat us. At least I hope they don't.

Of course, if they do, my job gets a lot easier. Bad astronomy would be the *least* of our worries.

22

❦❦❦❦❦❦

Hubble Trouble: Hubble Space Telescope Misconceptions

In 1946 astronomer Lyman Spitzer had a fairly silly idea: take a big telescope and put it in space. Looking back on his idea more than half a century later, it doesn't seem so crazy. After all, various nations have spent billions of dollars on telescopes in space, so *someone* must be taking the idea seriously. But in 1946 World War II was barely a year in the history books and the first launch of a satellite into orbit was still more than 10 years away.

Spitzer was a visionary. He knew that a telescope in space would have huge advantages over one on the ground, even before the first suborbital rocket flight gave others the idea that it was even possible. Sitting at the bottom of our soupy atmosphere yields a host of troubles for ground-based telescopes. The atmosphere is murky, dimming faint objects. It's turbulent, shaking the images of stars and galaxies until they all look like one blurry disk. Perhaps worst of all, our air is greedy and devours certain types of light. Some ultraviolet light from celestial objects can penetrate our atmosphere (the ultraviolet from the Sun is what gives us tans, or worse), but most of it gets absorbed on the way in. The same goes for infrared light, gamma rays, and x-rays. Superman may have x-ray vision, but even he couldn't see a bursting neutron star emit x-rays unless he flew up beyond the atmosphere, where there is no air to stop those energetic little photons.

I doubt Spitzer was thinking of Superman when he first proposed a space telescope, but the idea's the same. If you can loft a
telescope up, up, and away, out of the atmosphere, all those atmospheric problems disappear. The ultraviolet and other flavors of
light that cannot penetrate our atmosphere are easily seen when
you're above it. If the air is below instead of above you it won't
make stars twinkle, and the faint objects will appear brighter without the air glowing all around them, too.

Spitzer's vision became reality many times over. Dozens of telescopes have been launched into the Earth's orbit and beyond, but
by far the most famous is the Hubble Space Telescope (colloquially
called HST or just Hubble by astronomers). At an estimated total
cost of $6 billion, Hubble has made headlines over and over again.
Its images have made millions gasp in awe, and the astronomers
who use HST have learned more from it than perhaps any other
telescope in history, except, just maybe, Galileo's.

If you ask a random person in the street to name a telescope,
Hubble is almost certainly the only one he or she will know. However, sometimes the price of fame is misconception in the public
eye. Ask anything more specific, and that person will probably fal-
ter. Not many people know how big it is, where it is in space, or
even why it's in orbit. Some think it's the biggest telescope in the
world (or, more technically, *above* the world), some think it actually travels to the objects it observes, and others think it is hiding
secrets from the public.

At this point in the book you've figured out on your own that
none of these statements is true. Let's see why.

IT'S DONE WITH MIRRORS

Even the most basic aspects of the Hubble telescope are misunder-
stood. For example, CNN's web site, when describing one particular Hubble observation, had a headline that read, "Stars Burst into
Life before Hubble's Lens." Actually, Hubble doesn't have a lens.
Like most big telescopes, Hubble has a mirror that gathers and
focuses light. No less a luminary than Isaac Newton first figured
out that a mirror can be used instead of a lens, and the most basic

The Hubble Space Telescope floats freely over the Earth, prepared to take another observation of an astronomical object. Despite its clear views of the universe, the telescope is never more than a few hundred kilometers above the surface of Earth. (Image courtesy NASA and the Space Telescope Science Institute.)

design for a mirrored telescope is still called a Newtonian. Four hundred years later, it still causes confusion.

Lenses are good for smaller telescopes but become unwieldy when they are bigger than about a half-meter (20 inches) across. They have to be supported from the edges, lest you block their view. Large lenses are extremely heavy, which makes them difficult to use. They also need to be placed at the aperture of the telescope, at one end of a long tube. That placement makes the telescope unwieldy and very temperamental to balance.

Since only the front side of a mirror is needed, it can be supported all along its backside, making mirrors easier to use. A mirror reflects light, but a lens has to have light pass *through* it, which can dim that light. Even better, when you're making a mirror, you only need to grind and polish one side and not two. That's a pretty good savings over a lens.

Incidentally, the CNN web site made the same mistake less than a year later. I don't blame them, though. The Space Telescope Science Institute, which is charged with running the scientific aspects of Hubble, sponsored a PBS program about the telescope. In one episode, I heard an announcer introduce a segment as, "Through Hubble's Lens." If even PBS can get it wrong, what chance does everyone else have?

SIZE DOES MATTER

Many people are surprised at the large size of the Hubble satellite. It's roughly as big as a school bus. However, they are usually further surprised when told that the telescope is rather small as such things go. The primary mirror is 2.4 meters (8 feet) across. That may sound big compared to you or me, but there are many telescopes in the world more than four times that size. Even when Hubble was built it was not the biggest telescope. The legendary Hale Telescope at Pasadena's Palomar Observatory has a 5-meter (nearly 17-foot) mirror, and that one was built in 1936.

Not that Hubble is all that tiny. A full-scale mockup of it stands in a building at NASA's Goddard Space Flight Center in Greenbelt, Maryland. It's about five stories high and looms impressively over people as they walk by. Covered in shiny foil to reflect the Sun's light and keep the observatory cool, it looks like the world's largest TV dinner.

The reason Hubble isn't as big as some ground-based observatories is because it's hard to get something big into space. Hubble was designed to fit inside the Space Shuttle, and that put an upper limit on its size. The Next Generation Space Telescope, designed to observe infrared light and planned for launch in 2009, will be at least six meters (19 feet) across. One design calls for the mirror to be folded, and when it's out in space the mirror will unfold like a flower. Hubble's mirror, on the other hand, is basically one giant piece of glass, making it very heavy. If the mirror were any bigger, the spacecraft itself would have to be substantially larger to support it, making it impossible for the Space Shuttle to lift it.

A DROP IN THE BUCKET

A related misconception is that a telescope's most important function is to magnify an object, or make it look "closer." That's only partly true. It helps to make a small object look bigger, of course, but the real reason we make telescopes bigger is to collect more light. A telescope is like a rain bucket for light. If you are thirsty and want to collect rainwater, it's best to use a wide bucket. The wider the bucket you use, the more rain you collect. It's the same for telescopes: the bigger the mirror, the more light you collect from an object. The more light you gather, the fainter an object you can see. The unaided eye can pick out perhaps 10,000 stars without help, but with the use of even a modest telescope you can see millions. With a truly big telescope *billions* of stars become detectable.

The biggest telescopes on Earth have mirrors about 10 meters (33 feet) across, about the width of a small house. There are currently plans to build much larger telescopes. One design calls for a mirror 100 meters (109 yards) across! It's called the OWL, for Overwhelmingly Large Telescope. It'll cost a lot, but probably still less than Hubble did. A lot of that cost will probably go into simply finding a place to put it.

So Hubble may be small, but remember, it's above the atmosphere. The air glows, which washes out faint objects when viewed from the ground (see chapter 11, "Well, Well: The Difficulty of Daylight Star Sighting"). Hubble has darker skies and can see much fainter objects. The atmosphere also moves, so stars seen from the ground wiggle and dance (see chapter 9, "Twinkle, Twinkle, Little Star"). This spreads out the light from stars, making faint ones even more difficult to detect, especially if they are near brighter stars, which overwhelm them. With Hubble above the atmosphere, it avoids this effect and can more easily spot fainter stars. Between the much darker sky and ability to see faint objects, it holds the record for detecting the faintest objects ever seen: in a patch of sky called the Southern Deep Field, one of Hubble's cameras spotted objects *ten billion times fainter* than you can see with your unaided eye. That's a pretty good reason to loft a telescope a few hundred kilometers off the ground.

THE CASE FOR SPACE

Still, it's not easy getting something that size into space. For a long
time, Hubble was the largest single package delivered to orbit from
the Space Shuttle. The Shuttle can only get a few hundred kilometers above the Earth's surface, and schlepping the 12-ton Hubble
up made it even harder to get there. Using the Shuttle's robot arm,
in April 1990 astronaut Steve Hawley gently released Hubble into
the Earth's orbit, where it still resides, about 600 kilometers (375
miles) above the Earth's surface. It's another common misconception that Hubble is like the starship *Enterprise,* boldly going across
the universe to snap photos of objects no one has snapped before.
In reality, the distance from Hubble to the surface of the Earth is
about the same as that between Washington, D.C., and New York
City. Hubble is only marginally closer to the objects it observes
than you are! Sometimes it's actually *farther* from them; it may be
observing an object when it's on the far side of its orbit, adding a
few hundred kilometers to the distance the light travels from the
object to Hubble's mirror.

FILM AT 11:00

Which brings us to yet another common misthought about Hubble. Despite what many newspapers and television programs may
say, Hubble has never taken a single photograph of an object.
Hubble isn't a giant camera loaded with ISO 1,000,000 film. Hubble uses electronic detectors to take *images* of objects. These detectors are called **charge-coupled devices,** or **CCDs.** You've probably
seen or used one of these yourself: handheld video cameras have
been using CCDs for years, and digital cameras use them as well.
They are much better than film for astronomy because they are far
more sensitive to light, making it easier to detect faint objects.
They are stable, which means that an image taken with one can be
compared to another image taken years later. That comes in handy
when astronomers want to look for changes in an object's shape or
position over time. CCDs store data electronically, which means
the data can be converted to radio signals and beamed back to
Earth for processing. That's their single biggest advantage over film

for space telescopes. Who wants to go all the way into orbit just to change a roll of film?

PSST! CAN YOU KEEP A SECRET?

When Hubble points at an object, it's pretty likely to show us something we cannot see from the ground. That makes the data highly desirable, and of course that means a lot of competition to get time on the telescope. There aren't all that many astronomers around, but time on Hubble is an even rarer commodity. Once a year or so an announcement is made asking for proposals to use Hubble. Typically, NASA gets six times as many proposals as Hubble can physically observe during the upcoming year. Six-to-one odds are bit longer than most people like, but there is only so much observing time in a year. This creates a funny situation; a public telescope must, for a short time, have its data kept secret.

This time is called the **proprietary period,** and it is designed to give the astronomer a chance to look at the data. It may sound odd to keep Hubble data secret. After all, everyone's tax dollars paid for it, so shouldn't everyone have the right to see the data right away?

This may sound like a fair question, but really it's flawed reasoning. Your tax dollars pay for the IRS; then why not get access to your neighbors' tax returns? Ask the military for the blueprints for their latest secret fighter jet and see how far *that* gets you.

Now, to be fair, these really are secrets and there are good arguments for them to be. Hubble data are not really secret. But there is still good reason to let an astronomer have them for a year before they are released.

To see why, imagine for a moment you are an astronomer (if you *are* an astronomer, imagine for a moment you aren't so that you can then again imagine that you are). You have some nifty idea for an observation, and you decide you want to use Hubble for it. What do you do?

First, you'd better be sure you really need Hubble. Remember, for every astronomer trying to get time on Hubble, there are five others who are also vying for it, which means that right at the start you only have a one-in-six chance of getting your proposal accepted. That, in turn, means that the committee of astronomers that chooses who gets to use Hubble can be very picky. If your project can be done from the ground, you get rejected. If your project takes up too much time with only a marginal return in science learned, you get rejected. If you ask to do something that's already been done, you get rejected. If you ask to do something someone else is asking to do, and the other proposal is better, you get rejected.

Get the picture? It also takes days or weeks to prepare a proposal, time that you could spend working on other projects or trying to get other grants. You might use up a lot of precious time preparing your proposal only to have it roundly rejected.

But suppose you are lucky and your idea is accepted. Congratulations! Now you move to the next step. You have to painstakingly detail *every single thing* you want Hubble to do, including the initial pointing to your target, every exposure, every filter, every little bump and wiggle needed to get the observations you want. This detailing may also take several days or weeks, using complicated software guaranteed to give you a headache.

But finally you finish and submit the final proposal. Congratulations again!

Now you wait.

It may take up to a year or so to make those observations after the scheduling goes through. When you do, you are faced with many gigabytes of data, and you need a lot of software and experience to analyze them. It may take months or even years to figure everything out. With luck and perseverance, you may actually get a paper in the astronomical journals out of all this.

Now, think for a moment about all that work. All that analysis before and after the observations costs time and money, neither of which an astronomer has in copious amounts. For someone on a research grant time *is* money, and grants are very difficult to come by. Applying for HST time is a big gamble. You hope to get accepted, and then you hope the data are good enough to further your research, so you can get even more grants. I don't mean to put so much emphasis on money as a means unto itself, but without it, it's pretty hard to do research. In a sense, your future career as a scientist depends on your ability to get good data; you're staking your scientific reputation on your research. That Hubble data, once you've published it in an astronomical journal, is your lifeline.

Now imagine that the instant you get your data, some other astronomer has access to it, too. This other astronomer isn't as scrupulous and nice as you are. He or she also has experience with Hubble, knows just how to analyze your data, and might publish before you do! All that work, all that effort and time, and you get scooped with your own data.

That's why the data are held as proprietary for a year. That year gives the astronomer time to figure out what to do with the data and how best to analyze them. It's only fair to you, who devoted so much of your life to getting the data, to let you have a chance to look at them before anyone else.

So there's no real secret involved. At the end of the proprietary period, ready or not, the data become public. Far from being anything shady on the part of NASA, keeping the data secret for a year is actually the best way astronomers have come up with to further the cause of science in a fair manner. It can be an agonizing wait when you know some good data won't be available for a year, but it's worth it.

HUBBLE SHOOTS THE MOON

Hubble is more than just a telescope with a camera stuck onto it. It's a telescope with *several* cameras stuck onto it. Each instrument has a specific task. Some take ultraviolet images, others take infrared. Some take spectra by breaking the light from an object into individual colors. Each camera is a delicate, expensive piece of machinery.

Some of these instruments are very sensitive to light. They can actually be damaged if too much light hits them. Anyone who has ever had a roommate turn on a light in the middle of the night can sympathize with that.

This sensitivity has caused yet another myth about Hubble, that it cannot take images of the Moon. As the myth goes, the Moon is far too bright to be observed by Hubble without damaging these delicate instruments.

This turns out not to be the case.

It is true that the operators of Hubble need to be careful with subjecting the instruments to "an overlight condition." For example, there is a very strict "solar avoidance zone," a large area of sky around the Sun where Hubble is forbidden to look. The Sun *is* too bright, and if Hubble points too closely to it, the Sun would do all sorts of damage. That law is very stringently applied and has only been bent once, to observe the planet Venus.

However, this doesn't really apply to the Moon, which is far less bright than the Sun. While it's true that some of Hubble's cameras *are* very sensitive to light, they can simply be shut off during a lunar observation, allowing other, less sensitive cameras to be used. Still, many people have this mindset that you cannot look at bright objects, including the Moon. It's funny, because Hubble routinely observes the *Earth* and, from Hubble's vantage, the Earth is far brighter than the Moon.

The reason Hubble observes the Earth is nothing nefarious. Sometimes the great observatory is turned toward the Earth to take long exposures that help calibrate the cameras onboard. This allows astronomers to understand how the cameras behave. Hubble cannot easily track fast-moving objects, and the ground moves underneath Hubble at a clip of eight kilometers (five miles) per second. It makes a lousy spy satellite. The images are all streaked from the movement of the objects. I've seen some of these images, and you can clearly spot houses and trees that look like long, gray streaks. You don't have to bother shutting your window to protect your privacy. All Hubble sees of you is a long, blurry worm.

So, if Hubble can take images of the Earth, it certainly can take images of the Moon. The belief that the Moon is too bright is unfounded.

That said, why don't we see routine Hubble observations of the Moon?

For one thing, we already have really good images of the Moon from the Apollo missions and the Clementine lunar orbiter, better than Hubble can take. But there's more.

Here's where I sheepishly must admit to propagating my own little piece of bad astronomy. I'm commonly asked this question. I also *used* to say that Hubble *can't* take images of the Moon. It's not that the Moon is too bright, it's that it moves too fast. Hubble must be maneuverable enough to track nearby planets as they orbit the Sun, but the Moon moves across the sky much faster than even the fastest planet. There's no way, I would say, that it could track the Moon.

I was partially wrong in saying that. True, Hubble cannot track the Moon. But it doesn't have to track it. The Moon is *bright*. When you take an image of a bright object, you can take a shorter exposure. In fact, Hubble could take an image of the Moon with such a short exposure time that it would look as if the Moon were not moving at all. It's just like taking a picture out the window of a moving car. If you take a long exposure the trees will look blurred due to your motion. But if you snap a fast one the trees will look sharp and motionless. They don't have time to blur.

In 1999 just such an image of the Moon was taken by Hubble. The astronomers were clever. They put Hubble into "ambush mode," pointing it to a place where they knew the Moon would be and waiting for it to move into view. When it did, they took the image using a fast exposure. The results were pretty neat. They got nice pictures of the Moon, although not really any better than we had from orbiters. The principal goal of the observations was to get spectra of the lunar surface to help astronomers understand the properties of all the planets, and the images were an added bonus. So Hubble can indeed shoot the Moon, and did in the waning years of the twentieth century.

Ironically, while many people think that the Moon is too bright to observe with Hubble, it's the very brightness that allows Hubble to observe it! It's bright enough to let us take short snapshots of it without blurring.

TURNING THE CRANK

Unfortunately, the Moon issue won't die. Some people really want to see conspiracies and cover-ups everywhere they look, even when there are none to be had. One such person is Richard Hoagland, who maintains a long list of supposed NASA shenanigans, most of which involve space aliens. It would be fair to call Hoagland a kook. He leads the cause for the alien nature of the Face on Mars as well as a host of other fringe claims. On his web site (http://www.enterprisemission.com) he has an article about the Moon and Hubble with the headline: "NASA Caught in Yet Another Lie."

Hoagland quotes a Hubble astronomer and expert on Hubble imaging. As Hoagland relays on his web page, a UFO researcher asked the astronomer the following question: "Has Hubble taken any photos of the moon?" He responded: "No, the moon is too bright (even the dark side) to observe with HST."

I know this astronomer and called him about this. I swear I could hear his embarrassed smile on the phone. He apologized and said that the quote was sadly accurate, and he could kick himself for making the mistake. He simply wasn't thinking clearly and said the wrong thing. Unfortunately, with the web cranks can use his misstatement for their own ends. Hoagland claims that this is part of the NASA/alien cover-up of bases on the Moon. His headline about "NASA lies" is a bit disingenuous. A lie implies intent to deceive, while in reality an honest error was made. Also, the astronomer is not a NASA employee. Accuracy is perhaps not Hoagland's forte.

Hoagland's thrust is that this is just another NASA lie to cover up the fact that Hubble can indeed observe the Moon. According to his twisted logic, NASA spent years saying the Moon was off limits to Hubble to keep astronomers from finding the aliens. If that's true, why did NASA allow the team of astronomers to observe the Moon at all? In a rather typical example of conspiracy-theory logic, Hoagland ignores obvious facts that go against his conclusions.

It would be silly of NASA to maintain a conspiracy by claiming that the Moon is too bright to observe when, in fact, it was public record from the start that it routinely observes the much brighter Earth. Hoagland assumes, from one astronomer's single misstatement, that everyone in the astronomical community is part of a massive conspiracy and would blindly stick to an argument that is clearly contradictory to facts. Having worked with some of the astronomers and engineers who designed and use Hubble, I can assure you that these hardworking, intelligent, and clever people have no interest in covering anything up.

It gets even better. Not only was NASA *not* covering anything up, it actually initiated a fairly large amount of hoopla over the Hubble Moon images. Like most cranks, Hoagland is capable of weaving entire empires from fantasy, and would rather accuse people of lying than actually try to think logically for a moment.

In the end, the cranks and conspiracy theorists will believe whatever tale they tell themselves, as they always do and always will.

PAVED WITH GOOD INTENTIONS

I'll leave Hubble with one more story.

Probably my favorite media misadventure with Hubble involves that bastion of near-reality, the *Weekly World News*. Everyone knows their articles are jokes . . . or do they? It sells pretty well in grocery stores, and I always wonder how many people take it seriously. Headlines often scream, "Angels are Real—and Visiting Your Bathroom!" or "Boy Born Half-Bat Terrorizes Neighborhood!"

On July 19, 1994, the *News* had a story headlined "First Photos from Hell!" with the subtitle, "Listening device picks up screams coming from Black Hole!" (They use a lot of exclamation points.) According to the article, Hubble was observing a black hole when it detected a clear signal of people screaming. Obviously, these were the tortured souls of the damned in hell.

Ignoring for the moment (or forever) the silliness of Hubble picking up sounds at all, especially from hell, the best part of the article for me was the accompanying picture of a Hubble image of Supernova 1987a, a star that exploded in 1987. I studied this object for four years for my Ph.D., analyzing Hubble images and spectra. I sometimes worked until late at night trying to decipher what I saw, pounding my head on my computer screen in hopes of shaking loose some rusty cog in my brain. I never heard any tortured screams except my own.

So the last thing I need is for the *Weekly World News* to tell *me* that Hubble images of Supernova 1987a were hell. I wrote a whole thesis about it!

23

🐚🐚🐚🐚🐚🐚

Star Hustlers:
Star Naming for Dummies

When I was in high school, I had friend who was an expert on movies. He knew everything about every movie I had ever heard of. Director, actors, music, set design—the depth of his knowledge was amazing. One night at my house we were using my telescope and I said, "Let's take a look at Albireo. It's a cool double-star." I swung the telescope around and in a minute or two had it in the eyepiece. He stepped up to the eyepiece and took a moment to look at the pretty double. When he backed up, he took a look at the sky and said, "How in the world did you know where that star was? Look at all of them!"

I glanced up, and simply asked, "Who directed *From Here to Eternity?*"

Without missing a beat he replied, "Fred Zinnemann." He paused for a moment and then smiled. "Right," he said.

He understood. I knew the stars because I'm familiar with them. Reading the sky is like reading a map; after a while you know your way around. After you've seen a movie enough times, you get to know the characters, and if you're interested enough you'll learn details that not many other people know.

Decades later, I can make my daughter smile by pointing out stars to her. She wants to know their names, and I tell her. She repeats the name after me, but moves on to another star as quickly as she can. She wants to know *all* their names.

That's a tall order. There's no shortage of stars in the night sky. A keen-eyed observer—if the conditions are right—can see several thousand stars with the unaided eye. With even a modest telescope, hundreds of thousands of individual stars can be seen. The Hubble Space Telescope, in order to stay pointed at a target, employs a guide-star catalog that contains tens of millions of stars. As you'd imagine, naming them all can be quite a challenge.

But not for everyone. There are companies that offer to sell you the right to name a star after someone—yourself, perhaps, or a loved one or friend. For a fee, and not necessarily a small one, you receive a certificate authenticating some star in the heavens with the name you bestow on it. Some companies even give you the coordinates of your star and a stylish map so you can find it. There are many organizations like this, and one thing most have in common is that they strongly imply—and some come right out and say—that this star is now officially named after you. Congratulations!

But does that star *really* have your name? If you think so, I strongly urge you to close this book and read its title to yourself, out loud. Maybe twice.

The answer, of course, is no. The naming of stars is not a haphazard business. There is an organization called the International Astronomical Union that is in charge of giving celestial objects their official names. And by official, I mean the name that will commonly be used by professional astronomers when they refer to the object. There are rules for naming objects; asteroids, moon, comets, even craters on other planets get named in a certain way.

Stars typically have some sort of catalog name. As it happens, practically every star you can see with a modest telescope already *has* a name, or more properly a designation. Usually they are named for their position in the sky, which would be sort of like naming a tiny island after its longitude and latitude. Only the brightest ones, visible to the naked eye, might have proper names like Betelgeuse, Vega, or Polaris.

Most stars are named using Greek letters and the name of the constellation, like the famous Alpha Centauri or the not-so-famous Sigma Octans. The brightest star in the constellation is called Alpha, the second brightest is Beta, and so on. Those letters run out quickly, and so numbers are used after that. John Flamsteed was a seventeenth-century astronomer who catalogued thousands of stars, and many still bear his name. Over 300,000 fainter ones are listed in the German Bonner Durchmusterung catalog and bear the initials "BD" before a number representing their coordinates. Thousands of stars are in the Henry Draper catalog, named in honor of an astronomer who was among the first to use the new tool of spectroscopy in the 1870s (and who also took the first photograph of the Orion Nebula, 84 years to the day before I was born). These stars have the letters "HD" in front of a number representing *their* position on the sky.

Many stars are loaded down with a half-dozen or more obscure designations. Only a very rare few are named after individuals; van Maanen's star or Barnard's star are examples of those. These typically are special stars, like ones that are particularly close by or that have an unusually high velocity through the Galaxy. They're usually named after the astronomer who discovered their unusual properties. One star, Cor Coroli, is an exception—it's named after the heart of King Charles II, who patronized astronomy in the 1600s.

Not all of us are so lucky. Getting a star named after you is a very rare event.

Of course, the companies trying to sell stars would have you believe differently. You, too, can be immortalized in the heavens . . . if you believe their ads. Some are interesting indeed, claiming that astronomers will actually use the name you choose for the star. I'll let you in on a secret, as an astronomer: we don't. Many of us aren't particularly fond of the alphabet-soup names we use, but it's better than using the name "John Q. Public," and we don't have to change what we call a star because some company phones us to say that someone new has signed up for their "service."

The bottom line is, despite any claims by these companies, the name you give a star is just that: a name you give it. It isn't official and has no validity within the scientific community.

Now really, if all you care about is sending a unique gift to someone, and you like the fancy certificate, that's fine. But in their ads, many of these companies don't go out of their way to say that the names aren't *really* official. Many simply let you assume the name is official and do little to dissuade you of this notion.

Perhaps the most well-known star naming company is the International Star Registry (ISR). They claim to be the first company to sell star names, as if this gives them more of a foothold in the industry. Perhaps it does. Their web site claims they have sold hundreds of thousands of stars, and at $50 to $100 or so a pop you can do the math. The company isn't going broke.

They run a lot of ads on the radio. They used to claim that the star name you choose will go into a book in the Library of Congress and be printed in a book stored in a bank vault in Switzerland. In a sense, the former claim is true: any copyrighted material gets stored in the Library of Congress if it is registered by the claimant. The ISR is able to copyright their catalogs; a copyright is something you can buy on your own if you like. And if you have the cash to store a book in a Swiss vault, more power to you. This doesn't mean a whole lot as far as star names go, despite the ads.

So don't always believe what you hear. The New York City Office of Consumer Affairs certainly didn't. They levied a violation against the ISR for using deceptive advertising in New York City, with potential fines totaling up to $3,500 (a tiny fraction of the company's income). The Library of Congress pressured the ISR from citing the Library in the ISR's ads, and evidently they complied; the Library is no longer mentioned.

The astronomical community had something to say as well. You might think that astronomers wouldn't really care about this practice, since it doesn't directly affect them. Unfortunately, it can, and in a very emotional way. Consider this: Robert Martino, assistant director of the Perkins Observatory at Ohio Wesleyan University, points out that many people buy star names for friends or relatives who have died. He personally has had at least four groups of people at different times come to him and ask to see the star they named after their dead loved one. How does an astronomer tell a grieving person that the star doesn't really possess that name? Most astronomers don't; they point the telescope and swallow their anger.

Martino, however, finally reached his limit. He had faced too many grieving families, so he put up a scathing web page on the observatory web site about star naming. In the year 2000 the ISR retaliated.

According to Martino, the ISR put quite a bit of legal pressure on the observatory, which does not have a lot of money. Martino took down his page, although he was unhappy about it. Martino says nothing on his site was untrue. Just unflattering.

Martino also notes that the ISR was never directly indicated anywhere on his page. There was, however, a link at the bottom of the page about the New York City case, which did mention the ISR. Apparently, according to Martino, that was still too much for the company, which again contacted the university, warning them that the web site should not talk about star naming *at all*. The situation was quickly turning into one of First Amendment rights. Martino felt it was "a case of a consumer advocate being muzzled." According to Martino, after this event several astronomers who had web pages about star-naming companies edited them, prominently mentioning the First Amendment. Some sites even linked to a copy of the Constitution.

However, it didn't end there. Martino took down the web page but he was still incensed. He made his opinion clear on the Internet through various mailing lists and bulletin boards. Martino says the ISR once again contacted the university and insisted they wanted Martino to cease talking about them, claiming that Martino was representing himself as a spokesman for the university. This claim, Martino says, has "no basis whatsoever," and that his comments were made on his own time, using his private Internet account through his own Internet provider, and that the university had nothing to do with it. Still, the university sent Martino a letter making it clear that he'd better stop talking about them. Martino wound up moving the whole page about star naming to his private web site, where you can still find it at http://home.columbus.rr.com/starfaq.

Martino does extract some small amount of satisfaction, though. His new star-naming web page gets far more traffic than it did before the ISR contacted him. Evidently the publicity woke up other astronomers and they now link to his page as well.

I'll note that Martino has a daughter named Celeste: she is named after the stars and not the other way around.

❧ ❧ ❧

For their part, the ISR must be aware that many people buy stars as memorials; they have partnered with the Cancer Research Campaign, a company in the United Kingdom that raises money for cancer research. It's certainly understandable to do something to honor those who have died, especially family members. However, it might be better to donate money directly to a charitable organization, even more so if it's an organization promoting something about which you feel strongly.

Incidentally, on three separate occasions over the course of many weeks I called the ISR asking for comments on this situation, and even sent them a written letter. I also tried to get the university's side of the story. However, as of the time of this writing I have not received a reply from either of them.

The best thing to say is probably, "Caveat emptor." If you go in with your eyes open, understanding that star-naming is all completely unofficial, maybe there's no harm done. However, judging from stories I've heard from astronomers at planetaria and observatories, when most visitors ask to see "their" star, they don't understand that these companies are not official in any way. As the city of New York found, many of their ads really are deceptive.

❧ ❧ ❧

Ironically, the ISR's knowledge of astronomy could be better. The Australia-New Zealand office of the ISR has a web page (http://www.starregistry.com.au) where you can order a star name and find out more about the company. They have a "Frequently Asked Questions" page, and on it is the following gem:

QUESTION: What happens if my star falls out of the sky?
ANSWER: If this should happen, and came to our attention, we would most certainly name a new star for that person at our expense.

Usually, these FAQs are paraphrases of real questions, and it wouldn't surprise me if people asked this particular question. But a company that sells star names and makes all sorts of claims about astronomy should really understand the difference between a star in the sky and a shooting star, which is just another name for a meteor. Meteors have nothing at all to do with stars (see chapter 15 for more information about shooting stars). If an actual star fell out of the sky, we'd have bigger problems on our hands than finding a new star to hang a name on.

This same web site also claims there are 2,873 stars visible to the naked eye; in reality, there are more like 10,000 (depending on sky conditions). Besides being too small, that figure is awfully precise. How do they know it's not 2,872 stars, or 2,880? Using overly precise numbers sounds to me like another way to make them seem more scientific than they really are. If the ISR doesn't understand even the most basic properties of visual astronomy, do you really want to buy a star from them?

❧ ❧ ❧

Perhaps, after all this, it's time for me to come clean. I'll admit here that I have "my own" star. Many years ago my brothers bought it from the ISR and gave it to me as a birthday gift. That star—named Philip Cary Plait—is located in the constellation of Andromeda, and is about 100 times too faint to be seen with the unaided eye.

I lost the original certificate for the star years ago and out of curiosity I called the ISR to see if they could tell me where the star is. They were surprised; evidently, and ironically, it was one of the first stars sold by the company in their first year of business, but they were able to give me its coordinates. They were not very accurate but I was able to find it on a digital star map, which can be seen in the photograph.

Can you spot it? It's the one in the center. You can see there are many other stars in that field, including a lot that are brighter. None can be seen with the unaided eye, by the way. The kicker is that "my" star already has a name—BD+48° 683. For about 130 years this designation has been catalogued in the German Bonner Durchmusterung catalog used by practically every astronomer

The star "Philip Cary Plait"—a/k/a BD+48° 683—lurking not very
obviously in a field of thousands of other stars. The image shown is
1 degree across, roughly twice the size of the full Moon on the sky.
Image © 1995–2000 by the Association of Universities for Research in
Astronomy, Inc. The Digitized Sky Survey was produced at the Space
Telescope Science Institute under U.S. Government grant NAG W-2166.
(The images of these surveys are based on photographic data obtained
using the Oschin Schmidt Telescope on Palomar Mountain and the UK
Schmidt Telescope. The plates were processed into compressed digital
form with the permission of these institutions.)

on the planet. In the end, I think they have the edge in the naming
business.

So if you really want to buy a star, I urge you not to throw
your money at these companies. You could just go out and buy
some nice graphics software and make your own star-naming certificate, then pick any one you want, even the brightest in the night
sky, and it's just as official.

And I have an even better idea. Most observatories and planetaria are strapped for cash. Instead of buying a star, you could give them a donation to sponsor educational programs. That way, instead of just having *one* star you've never seen named for you, you'll be giving hundreds or thousands of people a chance to see *all* the stars in the sky.

Remember—the stars are for everyone, and they're free. Why not go to your local observatory and take a peek?

24

❧❧❧❧❧❧

Bad Astronomy Goes Hollywood: The Top-Ten Examples of Bad Astronomy in Major Motion Pictures

Whoosh! Our Hero's spaceship comes roaring out of a dense asteroid field, banks hard to the left, and dodges laser beams from the Dreaded Enemy, who have come from a distant galaxy to steal all of Earth's precious water. The Dreaded Enemy tries to escape Earth's gravity but is caught like a fly in amber. As stars flash by, Our Hero gets a lock on them and fires! A huge ball of light erupts, accompanied by an even faster expanding ring of material as the Dreaded Enemy's ship explodes. Yelling joyously, Our Hero flies across the disk of the full Moon, with the Sun just beyond.

❧❧❧

We've all seen this scene in any of a hundred interchangeable science-fiction movies. It sounds like an exciting scene. But what's wrong with this picture?

Well, everything, actually.

A lot of science-fiction movies are good fiction but *bad* science. Most writers have no problem sacrificing accuracy to make a good plot, and astronomy is usually the first field with its head on the block. How many times have you sat through one of these movies and shook your head at the way astronomy was portrayed? I spent a lot of my youth in front of the television watching bad science-fiction movies, and while they did help foment my interest in science, they also put a lot of junk into my head. So in the name of astronomers everywhere, I have compiled a Top-Ten list of Bad Astronomy in movies and TV. Some of these examples are specific and others are generalizations culled from hundreds of movies I've watched late at night or on Saturday mornings. The results are compiled into the scene above.

Let's pick apart that scene and find out just where it goes wrong. Go ahead and make some popcorn, sit back, drink some soda out of an oversized cup, and enjoy the show. And please! Be considerate of others; keep the noise to a minimum. Speaking of which . . .

1. *Whoosh! Our Hero's spaceship comes roaring out . . .*

Well, as they say, "In space, no one can hear you scream." Sound, unlike light, needs something through which to travel. What we hear as sound is actually a compression and expansion of the matter—usually air—through which the sound wave is traveling. In space, though, there's no air, so sound can't propagate.

But we live on a planet with a lot of air and we're accustomed to hearing things make noise as they go past us. Cars, trains, baseballs: they all whiz through the air as they pass us by. If we see something moving past quickly and quietly, it looks funny. Andre Bormanis, writer and science advisor for the *Star Trek* television series, confirmed a rumor I had heard for years: Gene Roddenberry, creator of *Star Trek,* wanted the original starship Enterprise moving silently through space. However, pressure from network executives forced him to add the familiar rumble and the "whoosh" as it flew past. In the later seasons, though, he removed the rumble. The "whoosh" in the opening credits stayed, though, probably (I'm guessing) because it would have cost too much to change the sequence. I guess the budget for space travel is tight even 200 years in the future.

There *is* a situation in which sound can propagate across space, when sound waves travel through an interstellar gas cloud. Even though they look thick and puffy, like the clouds after which they are named, a typical nebula (Latin for "cloud") is really not much more substantial than a vacuum. The atoms in the vast cloud are pretty far apart, but even a few atoms per cubic centimeter adds up when you're talking about a nebula trillions of kilometers thick. These atoms can indeed bump into each other, allowing sound to travel through the cloud.

Most processes that create "sound" in these nebulae, though, are pretty violent, such as when two clouds smash into each other or when a wind from a nearby star traveling at several kilometers per second slams into the nebula and compresses the gas. These processes generally try to push the gas around much faster than the nebula can react; the atoms of gas "communicate" with each other at the local speed of sound. If some atom is sitting around minding its own business and another one comes along moving faster than sound, the first atom is surprised by it. It's literally *shocked*: it didn't know what was coming. When this happens to a lot of material it's called a **shock wave**.

Shock waves are common in nebulae. They compress the gas into beautiful sheets and filaments, which we can "ooohh" and "ahhh" at from our nice comfortable planet safely located a few hundred light-years away. I imagine property values near the Orion Nebula are at a premium. The view is unparalleled, and if you choose your site correctly the ghostly whispers of swept-up atoms will remain unheard.

2. . . . of a dense asteroid field . . .

Ever heard the term "asteroid swarm"? Well, it's more like an "asteroid vacuum." In our solar system the vast majority of asteroids are located in a region between Mars and Jupiter. The total amount of area defined by the circles of their two orbits is about one-quintillion (10^{18}) square kilometers. That's a lot of room! Astronomer Dan Durda puts it this way: imagine a scale model of the solar system where the Sun is a largish beach ball a meter (1 yard) across. The Earth would be a marble 1 centimeter ($1/2$ inch) in size located about 100 meters (roughly the length of a football field) from the Sun. Mars would be a pea about 150 meters away from the Sun, and Jupiter, the size of a softball, about 500 meters out.

If you collected all the asteroids in the main belt and balled them up, they would be *in toto* about the size of a grain of sand. Now imagine crushing that grain of sand into millions of pieces and strewing it over the hundreds of thousands of square meters between Mars and Jupiter in the model. See the problem? You could tool around out there for months and never see an asteroid, let alone two.

In *The Empire Strikes Back* Han Solo has to do some pretty tricky maneuvering in an asteroid field to avoid being turned into Smuggler Paste by the Imperial starships. Those rocks were pretty big, too, dwarfing the Millennium Falcon. Let's say the average asteroid in that swarm was 100 meters across, and the average distance between them was 1 kilometer (0.6 miles)—we're being very generous here! Given the average density of rock (a couple of grams per cubic centimeter), that would give each asteroid a mass of about a trillion grams, or about a million tons. That in turn means the entire swarm, if it is the same size as our own asteroid belt, would have a mass of about 10^{30} grams. That's about a million times the mass of our own asteroid belt, or the combined mass of all the planets in our solar system. That's one big asteroid swarm. No wonder Solo could hide his ship there!

It's possible that in other solar systems, asteroid belts are bigger. We have just started detecting planets orbiting other stars, and these exosolar systems are very different than our own; we have just the beginnings of a cosmic diversity program. We don't have the technology yet to know what the asteroid belts in these other systems look like or if they even *have* asteroid belts. Still, a lot of movies use a very dense asteroid "storm" to advance the plot. (The original TV series *Lost in Space* used one to throw the Jupiter 2 off course, and *Star Trek* used it as an excuse to damage a vessel so that it could be rescued by Kirk and crew.) How many of them can there be? I suppose we'll just have to wait and see.

3. . . . *banks hard to the left* . . .

Once again we run into a lack of air up there. We moribund humans are conditioned to expect airplanes to bank as they make turns. Tilting the wings of the plane helps redirect the thrust to the side, turning the plane. But note what is doing the pushing: air. Need I say it? No air in space.

To make a turn in space, you need to fire a rocket in the opposite direction that you want to turn. Need to escape to port? Thrust starboard. Actually, banking makes the situation even worse: it presents a broader target to a pursuing enemy. Keeping the wings level means less ship to aim at. Speaking of which, why do so many movies have spaceships with wings in the first place?

To be fair, I'll note that banking has one advantage. When a car makes a turn to the left, the passengers feel a force to the right. That's called the **centripetal force,** and it would work on a spaceship, too. Extensive tests by the Air Force have shown that the human body reacts poorly to high levels of acceleration. A seated pilot accelerated upward experiences forces draining blood away from the brain, blacking him out. If he's accelerated downward, blood is forced into the head, an unpleasant feeling as well. The best way for the body to take a force is straight back, pushing the pilot into his or her seat. So, if a pilot flying a spaceship banks during a turn, the centripetal force is directed back, pushing the pilot harder against the seat. Blacking out during a space battle is not such a hot idea, so maybe there's some truth to banking in space after all.

One other thing: if the spaceship has artificial gravity, then the computer should be able to account for and counteract any centripetal force. So if you see a movie in which Our Heroes have gravity onboard and still bank, you know that you're seeing more bad astronomy.

4. . . . and dodges laser beams from the Dreaded Enemy . . .

If screenwriters have a hard time with the speed of sound, imagine how difficult it must be for them to work with the speed of *light*. Perhaps you've heard the phrase "300,000 kilometers (186,000 miles) per second: not only a good idea, it's THE LAW!" They aren't kidding. According to everything we understand about physics today, nothing can travel faster than light. Now I accept that someday, perhaps, we may find a way around that limit. No one wants to do that more than astronomers: they would give up their biggest grant to climb aboard a spaceship and zip around the

Galaxy. To be able to actually *see* a planetary nebula from up close, or to watch the final seconds as two madly whirling neutron stars coalesce in an Einsteinian dance of mutual gravitation: that's why we went into astronomy in the first place! But right now, today, we know of no way to travel or even to transmit information faster than light.

Therein lies the problem. Laser beams travel at the speed of light, so there is literally no way to tell that one is headed your way. There's more: out in space, you can't see lasers at all. A laser is a tightly focused beam of light, and that means all the photons are headed in one direction. They go forward, not sideways, so you can't see the beam. It's just like using a flashlight in clear air: you can't see the beam, you only see the spot of light when it hits a wall. If you see the beam, it's because stuff in the air like particles of dust, haze, or water droplets is scattering the photons in the beam sideways. In laser demonstrations on TV you can see the beam because the person running the demo has put something in the air to scatter the beam. My favorite was always chalk dust, but then I like banging erasers together. Anyway, if you're in a laser battle in your spaceship, you really won't see the enemy shot until it hits you. Poof! You're space vapor (ironically, a second shot fired *would* get lit up by all the dust from your exploding ship). Sorry, but dodging a laser is like trying to avoid taxes. You can try, but they'll catch up to you eventually. And unlike lasers, the IRS won't be beaming when it finds you . . .

5. . . . *who have come from a distant galaxy* . . .

Even the awesome speed of light can be pitifully dwarfed by the distances between stars. The nearest stars are years away at light speed, and the farthest stars you can see with your naked eye are hundreds or even thousands of light-years away. The Milky Way Galaxy is an unimaginably immense wheel of hundreds of billions of stars, over one-hundred-thousand light-years across—

—which in turn is dwarfed by the distance to the Andromeda galaxy, the nearest spiral galaxy like our own. M31, as astronomers in the know call it, is nearly *three million* light-years away. Light that left M31 as you look at it in your spring sky started its journey when *Australopithecus afarensis* was the most intelligent primate on the planet. And that's the *nearest* spiral. Most galaxies you can see with a modest telescope are a hundred-million light-years away or more.

Now, doesn't it seem faintly ridiculous for aliens to travel from some distant galaxy to the Earth? After all, the distances are pretty fierce, and they have many, many stars to plunder and pillage in their own backyard. Science-fiction movie writers tend to confuse "galaxy," "universe," and "star" quite a bit. The 1997 NBC made-for-TV movie, *Invasion,* was advertised as having aliens travel "over a million miles" to get here. Ironically, ad writers wanted that distance to sound huge, but consider this: the Moon is only a quarter of a million miles away, and the nearest planet about 25 million miles away. The nearest star to the Sun, Alpha Centauri, is 26 million-*million* miles away. It sounds like they grossly underestimated the size of the gas tanks on the alien ships.

6. . . . to steal all of Earth's precious water . . .

This is my personal favorite. It was used in the 1980s TV movie, *V,* and countless other pulp sci-fi movies. This may have started in the late 1800s, when astronomer Percival Lowell thought he saw canals on Mars and concluded that the planet was drying up. Obviously, an advanced race was trying to save itself via irrigation. Unfortunately, what he really saw were faint features on Mars that his all-too-human brain tried to connect up in his imagination. There are no canals on Mars.

On the face of it, that aliens want our water seems plausible: look at all the water we have on Earth. Our planet is three-quarters covered in it! Desperate for water, what would our proposed aliens do? After looking toward our blue world with envious eyes and parched tongues (or whatever they had in their mouths, if they even *had* mouths), would they come all the way in to the center of the solar system, using up huge amounts of energy to get in and out of the steepest part of the Sun's and Earth's gravity wells, to suck up water in its very inconvenient liquid form?

No way. Water is *everywhere* in the solar system. Every outer moon in our system has quite a bit of frozen water. Saturn's rings are mostly composed of water ice. And if that's no good, there are trillions of chunks of ice prowling the cold vastness of the Oort cloud, the cometary halo of the Sun that is almost a light-year across. Why expend all that energy to get to Earth when you can mine the ice out of all those comets, a trillion kilometers from the heat and fierce gravity of the Sun? And ice is a very convenient form of water. It may take up slightly more room than liquid water, but it doesn't need a container. Simply chisel it into the shape you want, strap it to the outside of your ship, and off you go.

Of course, in *V,* besides stealing our water, the aliens also came here to eat us. In that case, they *did* have a good reason to come to the Earth. Tough luck for us. Still, if I were some ravenous alien with a taste for human flesh, I'd simply gather up a bunch of cells and clone them to my heart's (or whatever) content. Why travel hundreds of light-years to eat out when staying home is so much easier?

7. The Dreaded Enemy tries to escape Earth's gravity, but is caught like a fly in amber.

How many times have you heard the phrase, "escape from Earth's gravity"? Technically it's impossible. According to Einstein, the mass of the Earth bends space, and the farther away you get, the less space gets bent. We feel that bending as gravity. But even Albert would agree with Isaac Newton that in general terms, the force you feel from gravity weakens proportionally as the square of the distance. So, if you double your distance from the Earth, you feel a force one-quarter what you did before. If you go 10 times farther away, that force drops by a factor of 100. You'll note that gravity drops off fast, but not *infinitely* fast. In other words, even if you go a billion times farther away, you will still feel some (extraordinarily small) force. Gravity never goes away, and if you forget that for an instant you'll be sorry. Toddlers tend to learn it pretty quickly.

So if gravity is always around, it's not like you are floating care-free one instant and suddenly feeling a strong gravitational force the next. It's a gradual change as you approach an object. *Star*

Trek would sometimes have the Enterprise lurch as it approached a planet and gotten "stuck" in the gravity, sending hapless crewmembers flying from their stations. Luckily, the universe doesn't behave that way.

You'd think after the second or third time that happened, someone down in the Enterprise's engineering section would have whipped up some seat belts.

8. As stars flash by . . .

When you're talking real estate in outer space, it's not location, location, location but scale, scale, scale. Planets are pretty far apart, but stars are really, really, *really* far apart. The nearest star to the Earth (besides the Sun) is about 40 trillion kilometers (25 trillion miles) away. Even distant Pluto is 8,000 times closer than that. You can go all the way across our solar system and, to the naked eye, the stars will not have appeared to move at all. The constellations will look the same on any planet in the solar system.

But actually, if you go to Pluto, for instance, the stars *will* appear to move a tiny but measurable amount. The European satellite Hipparcos was launched specifically to measure the change in the apparent position of stars as it orbits the Earth. By making exact position measurements, you can determine the distance to nearby stars. Hipparcos has already revolutionized our ideas on the size of the universe simply by finding that some stars are about 10 percent farther away than previously thought. The downside of this, of course, is that the commute for the aliens is longer.

I was once fooled by someone asking what was the nearest star to the Earth. "Proxima Centauri!" I piped up, but of course the real answer is the Sun. In the movie, *Star Trek IV: The Voyage Home,* the Enterprise and crew need to warp past the Sun to go back in time. There are two problems with this scene. One is that you can actually see stars moving past them as they travel to the Sun; there aren't any. Second, at the speed of light, the Sun is a mere 8 minutes away. At warp 9 they would have zipped past the Sun in less than a second. That would have made for a short scene.

9. . . . Our Hero gets a lock on them and fires! A huge ball of expanding light erupts past us, accompanied by an even faster expanding ring of material as the Dreaded Enemy's engines explode.

Explosions in space are tricky. Stuck here as we are on the Earth, we expect to see a mushroom cloud caused by the superheated air in the explosion rising rapidly, accompanied by an expanding circle of compressed air formed by the shock wave as it moves along the ground.

The lack of air in space strikes once again. In the vacuum of space there is nothing to get compressed. The expanding shell of light that is the trademark of most science-fiction explosions is just another way to make viewers feel more at home. The debris itself expands more slowly; pieces fly out in all directions. Since there is no up or down in space, the explosion will tend to expand in a sphere. The debris will no doubt be very hot, so we might actually see what looks like sparks exploding outward, but that's about it.

Of course, it's a lot more dramatic to have nifty things happen during an explosion. The quickly expanding shell of light looks really cool, if implausible. Sometimes, though, it makes some sense. In the movie, *2010: the Year We Make Contact,* Jupiter is compressed by advanced alien machinery until it is dense enough to sustain nuclear fusion in its core. The core ignites, sending a huge shock wave through the outer atmosphere. This would get blown off and be seen as an expanding shell of light. That was relatively accurate and fun to watch, besides.

A special effect tacked on in recent movies is the expanding ring of material seen in explosions. This started with *Star Trek VI: The Undiscovered Country,* when Praxis, the Klingon moon, exploded. The expanding ring that results is for my money the most dramatic effect ever filmed. I also have to give this scene the benefit of a doubt. The expanding ring we see during a large explosion on Earth is shaped by the ground itself. You can think of it as part of the explosion trying to move straight down but being deflected sideways by the ground. In space, you wouldn't get this ring, you'd get a sphere. But the explosion in *Star Trek VI* was not a simple one; it's possible the expansion was distorted by the shape of the moon. A flat ring is unlikely but not impossible.

In the special edition of *Star Wars: A New Hope,* released in 1997, the Death Star explosion at the end (hope I didn't spoil it for you) also features an expanding ring. Once again, I'll defend the effect: explosions, like electricity, seek the path of least resistance. Remember, the Death Star had a trench going around its equator. An explosion eating its way out from the center would hit that trench first and suddenly find all resistance to expansion gone. Kaboom! Expanding ring.

We see expanding rings in real astronomy as well. The ring around Supernova 1987a is a prime example. It existed for thousands of years before the star exploded, the result of expanding gas being shaped by gas already in existence around the star. Even though not technically caused by an explosion, it shows that sometimes art imitates nature.

10. *Yelling joyously, Our Hero flies across the disk of the full Moon, with the Sun just beyond.*

The phases of the Moon always seem to baffle movie makers. The phase is the outcome of simple geometry: the Moon is a sphere that reflects sunlight. If the Sun is behind us, we see the entire hemisphere of the Moon facing us lit up, and we call it a full Moon. If the Sun is on the other side of the Moon, we see only the dark hemisphere and we call it a new Moon. If the Sun is off at 90 degrees from the Moon, we see one-half of the near hemisphere lit, and we call it half full or, confusingly, a quarter moon, since this happens one-quarter of the way through the Moon's phase cycle. This is explained in detail in chapter 6, "Phase the Nation."

In the 1976 British television program *Space: 1999,* for example, the Moon is blasted from Earth's orbit by a bizarre explosion (which in itself would be bad astronomy but is later explained in the series to have involved an alien influence). In the show, we would always see the Moon traveling through deep space in a nearly full phase. Just where was that light coming from? Of course, in deep space there is no light source, which would have made for a pretty boring shot of the Moon.

Even worse, in movies and a lot of children's books the Moon is sometimes depicted with a star between the horns of the crescent.

That would mean a star is between the the Moon and the Earth. Better grab your suntan lotion!

❧ ❧ ❧

Our fictionalized movie scene has some dreadful astronomy in it, and we haven't even touched on black holes, star birth, and what nebulae really look like. But what movies have *good* astronomy? Any astronomer will instantly reply: *2001: A Space Odyssey.* In that movie, for example, the spaceship moves silently through space (a fact they evidently forgot when making the sequel *2010: the Year We Make Contact*). There are countless other examples. An astronomer once told me that the only mistake in the movie is when one of the characters, on his way to the Moon on the PanAm shuttle, takes a drink from his meal and you can see the liquid in his straw go back down after he finishes sipping. Since there is no gravity on the shuttle, the liquid would stay drawn up in the straw. This is nit-picking at an almost unbelievable level, and I think we can forgive the director.

Surprisingly, the TV show *The Simpsons* commonly has correct astronomy. There is an episode in which a comet threatens to collide with the Earth. The comet is shown being discovered by an amateur (our antihero, Bart). Most comets are indeed discovered by amateurs and not professionals. Bart then calls the observatory to confirm it, which is also the correct procedure (he even gives coordinates using the correct jargon). When it enters the Earth's atmosphere, the comet is disintegrated by all the smog in the air of the Simpsons' overdeveloped city. That part can be chalked up to comedic license, but then comes an extraordinary scene: The part of the comet that gets through the pollution is only about the size of "a Chihuahua's head," and when it hits the ground, Bart simply picks it up and puts it in his pocket. As we saw in chapter 15, contrary to common belief, most of the time a small meteorite will not be burning hot when it hits the ground. The rock (or metal) is initially moving very rapidly through the upper atmosphere, which will melt the outer layers, but friction very quickly slows the rock down. The melted parts get blown off and the remaining chunk will only be warm to the touch after impact. In this episode of *The*

Simpsons, they imply that the comet chunk is hot but not too hot to pick up. That's close enough for me.

After the original version of this chapter appeared in *Astronomy* magazine in April 1998, I received a letter from a young girl accusing me of ruining science-fiction movies for her. I have also received the occasional e-mail from my web site, where I review specific movies like *Armageddon, Deep Impact,* and *Contact,* telling me to either "get a life" or "learn how to just enjoy a movie." On the other hand, I get a hundred times as much e-mail agreeing with my reviews. Still, dissenters have a valid point. Do I really hate Hollywood movies?

Armageddon notwithstanding, no I don't. I like science fiction! I still see every sci-fi movie that comes out. When I was a kid I saw just about every science fiction movie ever made. I ate up every frame of rocket ships, alien monsters, evil goo, and extraterrestrial planets, no matter how ridiculous or just plain dumb the plot.

So what's the harm? You may be surprised to know that I think it is minimal. Although bad science in movies does reinforce the public's misunderstanding of science, the fact that science fiction does so well at the box office is heartening. Most of the top-ten movies of all time are science fiction, showing that people really do like science in movies, even if it's, well, *bad.* I would of course prefer that movies portray science (and scientists!) more realistically. Sometimes science must be sacrificed for the plot, but many times, maybe even most of the time, correct science could actually improve the plot. Thoughtful movies do well, too, like *Contact* and, of course, *2001,* now a classic of science fiction.

If movies spark an interest in science in some kid somewhere, then that's wonderful. Even a bad movie might make a kid stop and look at a science book in the library, or want to read more about lasers, or asteroids, or the real possibility of alien life. Who knows where that might lead?

For me, it led to a life of astronomy. I can only hope that even bad astronomy, somehow, can spark good astronomy somewhere.

RECOMMENDED READING

No book on astronomy could possibly cover every topic in every detail without stretching from here to the Moon and maybe even back again. The following list represents just a few books and web sites that might help you pursue the topics covered in this book a bit further. Many of them helped me a great deal when researching *Bad Astronomy*.

Books

Carl Sagan did so much for public outreach in astronomy and science that scientists everywhere owe him an enormous debt. Of his many works, by far the finest—and the most fun to read—is *The Demon Haunted World: Science as a Candle in the Dark* (Ballantine Books, 1997, ISBN 0-345-40946-9). It's a brilliant look at skepticism in many disciplines, and can be easily be applied to everyday life outside the observatory.

Stephen Maran has also helped the public understand astronomy for many years. His book *Astronomy for Dummies* (IDG Books Worldwide, 2000, ISBN 0-7645-5155-8) is a fun and helpful guide to the universe.

I turned to Joel Achenbach's *Captured by Aliens: The Search for Life and Truth in a Very Large Universe* (Simon & Schuster, 1999, ISBN 0-684-84856-2) expecting to read a silly exposé of people who think they are channeling aliens from another dimension, but instead found a thoughtful but still funny book about people trying to cope with modern times.

John Lewis's *Rain of Iron and Ice* (Helix Books, 1996, ISBN 0-201-48950-3) is a fascinating look at asteroid and comet impacts. It's riveting, and might scare you a little. I have always said that no one has ever been documented to have been killed by a meteor impact . . . but that was *before* I read this book.

In this short book I could only scratch the surface of the Velikovsky affair. Numerous books have been written about it, but you can start

259

with the man himself: Immanual Velikovsky's *Worlds in Collision* (Doubleday, 1950). I also recommend the transcripts of the AAAS debate mentioned in the text *Scientists Confront Velikovsky*, edited by Donald Goldsmith (Cornell University Press, 1977).

In the end, one of astronomy's most rewarding gifts is simply the stunning beauty of the universe. There are many wonderful astronomy books loaded with great pictures; a recent and very good one is astronomer Mark Voit's *Hubble Space Telescope: New Views of the Universe* (Harry N. Abrams, with the Smithsonian Institution and the Space Telescope Science Institute, 2000, ISBN 0-8109-2923-6). This coffee-table book will have you thumbing through it again and again, staring in amazement at the glorious pictures.

The World Wide Web

Or as I like to call it, "The Web of a Million Lies." For every good astronomy site, there seem to be a million that are, uh, *not* so good. But if you have a guide and a skeptical eye, there are a lot of web sites out there that will sate your thirst for astronomical knowledge. If these don't do it, you can always try your favorite search engine. But knowing the web as I do, you might want to search it with both eyes in a permanent squint. Maybe it would be better to just go with the sites listed below.

If I may be so immodest, I'll start with my own: Bad Astronomy (http://www.badastronomy.com). You'll find a few of the same topics covered in this book and many other as well. There are also links to other sites that will keep you busy for a long, long time. (Believe me.)

Penn State University meteorologist Alistair Fraser's Bad Science web site (http://www.ems.psu.edu/~fraser/BadScience.html) was in many ways the inspiration for my own. A weather kind of guy, Alistair has created a site that is a bit more down to Earth than my own.

Bakersfield College astronomer Nick Strobel has put together a wonderful web site called Astronomy Notes (http://www.astronomynotes.com), which covers everything from navigating the night sky to the shape and fate of the universe. I rely on it quite a bit to help explain why things happen the way they do.

Bill Arnett is not a professional astronomer, but he fooled me into thinking so. His Nine Planets web site (http://seds.lpl.arizona.edu/nineplanets/nineplanets/nineplanets.html) is an amazingly complete and informative place to find out just about everything you want to know about the solar system. Each planet gets its own page as do some moons, and he has a huge list of links to pictures on every page.

Mikolaj Sawicki is a physicist at John A. Logan College in Illinois. His web site on tides (http://www.jal.cc.il.us/~mikolajsawicki/gravity_and_tides .html) cleared up some of my own tidal misconceptions. It has a very clear and interesting explanation of tides, and is one of the very few that not only is correct but carries out the idea to its logical conclusions.

One of the great aspects of the web is the amazing amount of information it contains—sometimes it's even accurate. So many questions come up so frequently that people often put together Frequently Asked Questions lists, or FAQs. The Astronomy FAQ (http://sciastro.astronomy .net/) may, then, answer many of your questions. The Physics and Relativity FAQs (http://math.ucr.edu/home/baez/physics/) do the same for their fields and would please Uncle Albert himself. Each of these FAQs has links to even more web sites, which keep even a hardened geek like me busy for hours on end.

If that's not enough, try astronomer Sten Odenwald's Ask the Astronomer web page (http://itss.raytheon.com/cafc/qadir/qanda .html). He has answered over 3,000 questions, so any you have might already be there.

Once again, if pictures are what you're after, then try either the Space Telescope Science Institute's web site (http://www.stsci.edu) or the amazing Astronomy Picture of the Day (http://antwrp.gsfc.nasa.gov), which, true to its name, has a new beautiful picture posted each day. These are two of the most popular sites on the web, in any topic, and it's not hard to see why.

While researching the chapter on the Apollo Moon Hoax, and later when looking for images and information about Apollo, I turned again and again to the *Apollo Lunar Surface Journal* at http://www.hq.nasa.gov/ office/pao/History/alsj/. There you will find an astonishing amount of detail about the most ambitious and successful space adventure in human history. I fell in love with space travel all over again after going through the images there.

There are a lot of great web sites promoting skepticism in general. I highly recommend the Talk Origins Archive (http://www .talkorigins.org), which is a pro-science web site that is mostly an answer to creationist arguments. It leans heavily toward evolution, but has great astronomy pages, too.

There are a number of web sites devoted to Immanual Velikovsky's ideas, both pro and con. The biggest one on his side is http://www .varchive.org, which has many of his writings. A good web site debunking Velikovsky is the Antidote to Velikovskian Delusions at http://abob .libs.uga.edu/bobk/velidelu.html.

One of the most wonderfully rational and skeptical sites on the web is run by none other than James Randi, the Amazing Randi himself. Randi has devoted his life to debunking pseudoscience and paranormal claims, and does so in a tremendously entertaining way. His web site (http://www .randi.org) is a vast store of rational treasures, from his now-famous $1 million challenge for proof of the paranormal to his essays railing against fuzzy thinking.

Finally, if you're an aficionado of bad movies, as I am, try the Stomp Tokyo Video Reviews (http://stomptokyo.com), a loving, and sometimes not-so-loving, look at B movies. These guys really need to get out more, but I love their site.

ACKNOWLEDGMENTS

I feel obligated as a first-time author to thank every person I have ever met. I have been accused of being overly wordy (mostly by my editor at Wiley, Jeff Golick). I cannot imagine what he and others are thinking, but to give them the benefit of the doubt, I'll be brief.

Lynette Scafidi is a naturalist at the Brookside Nature Center in Wheaton, Maryland. Through a coincidence set in motion by her, I became acquainted with Steve Maran, a well-known astronomy author and press officer for the American Astronomical Society. Steve helped my public outreach career tremendously, and he also recommended his own literary agent to me when I asked about writing a book. Skip Barker is a pretty good agent to a punk astronomer trying to break into the business, and he did a lot of legwork in getting this book published. I owe the basic existence of this book to Lynette, Steve, and Skip. Speaking of Lynette, I strongly recommend a side-trip to the Brookside Nature Center for anyone visiting Washington, D.C. Call them at (301) 946-9071.

My thanks goes to Dr. Mark Voit, who agreed to be the technical editor for this book. Mark is a professional astronomer who does a lot of work with Hubble and really knows his stuff. If there are any remaining technical errors in this book, blame me and not him.

There are a large number of people who helped guide my thinking and actions while preparing this book. Among them are Paul Lowman, Alistair Fraser, Ken Croswell, C. Leroy Ellenberger, Mikolas Sawicki, Dan Durda, Bill Dalton, Delee Smith, Barb Thompson, Rebecca Eliot, and of course the pack of weirdos who frequent the Bad Astronomy Bulletin Board on my web site. Also, thanks to the people at *Astronomy* magazine for allowing me to use an unedited version of my April 1998 article "A Full Moon in Every Plot," which in full bloom is in this book as chapter 24.

Thanks to my boss, Lynn Cominsky, for talking up my site to people who could influence my career, and to the people with whom I work, for not getting bored listening to me rant about this stuff. Also to Dan

Vergano, for giving me a heads-up on the Apollo hoax when I really needed it.

A special thanks to my friend Kat Rasmussen for her great work in turning my Bad Astronomy web page from a tiny, chaotic mess into the lumbering juggernaut it is today. Check out her stuff at http://www .katworks.com.

Jeff Golick did indeed cut my precious prose when I became too long in my word usagement. I have a hard time remembering I am not paid by the word. Anyway, his suggestions were all quite good, except for the one about putting punctuation inside the "quotation marks".

Of course, I thank my family for supporting me in my obsession with astronomy all these years. Especially my Mom and Dad, who bought a cheap department-store telescope and let their four-year-old kid look at Saturn through it. It changed my life forever. Three decades (plus some) later, their act led to the book you hold in your hands. Expose your kids to science whenever you get the chance. You never know where it will lead.

My second biggest thank-you goes to my in-house editor, Marcella Setter, who, unimaginably to me, also let me marry her. She dots my eyes and crosses my tees, and without her I would be lost in a Bad Universe.

But most of all, to Zoe. You're the reason I'm doing this in the first place. I love you.

INDEX